Multiplication is Fun

The Lifematics

The Lifematics Centre
Telephone: 08068464607, 08078611408
E-mail: thelifematics@gmail.com, contact@thelifematics.com.ng
Website: www.thelifematics.com.ng

ISBN 978-978-947-288-8

Acknowledgments

The Lifematics would like to acknowledge the support of those who demonstrated their commitment to the book.

We appreciate the efforts of Moyo Jogunosimi, the Lifematics's founder who worked tirelessly on the book. We would also like to recognise the support of the Lifematics team.

We are grateful to Eyo Inameti for the photographs. We are also very grateful to all our students who were the first set of people we taught these techniques and Toluwani Toyin-Gbede for helping out with some of the typing.

The Lifematics,
2015

Preface

Of all the four major arithmetic operations; addition, subtraction, multiplication and division, multiplication is required of every student to memorise at some point in their academic endeavour. Many children spend hours memorising and how unhappy they are, when they make mistakes while reciting. If tips are given on multiplication, it surely will be more enjoyable.

The purpose of this book is to provide different techniques of multiplying. Though learning all the techniques is great, readers are advised to stick to the technique they are most comfortable with. However, do not get discouraged if you do not master it quickly; always remember that it is constant practise that brings perfection.

Should you know other multiplication techniques, do well to send a mail to: thelifematics@gmail.com or contact@thelifematics.com.ng

Contents

Glossary

Commutative: giving same result irrespective of the order of the numbers.

Digit: a symbol that represents numbers.

Dividend: a number to be divided by another number.

Divisor: a number that is used to divide.

Denominator: the number below or to the right of the line in a fraction. It shows how many equal parts the item is divided into.

Diagonal: a straight line that connects two opposite corners together.

Even Number: a number that can be divided by 2 with no remainder.

Horizontal: a line that is parallel to the floor that is drawn from left to right or right to left.

Matrix: an organised set of numbers in rows and columns.

Mental: relating to the mind.

Monetary: relating to money or currency

Numerator: the number on top or to the left of the line in a fraction.

Odd number: a number that has remainder 1 after it is divided by 2.

Place-Value: the value is determined by the position of the digit within a number.

Successive: following a uniform pattern.

Times Table: a list that shows the result of numbers mutiplied.

Intersection: a point where two lines cross each other.

Vertical: a line that stands upright that is, drawn from up to down or down to up.

Acronyms

DIY: Do It Yourself

E.G: Example

I.E.: That is

LHS: Left Hand Side

RHS: Right Hand Side

Math operation family

The four major Mathematical operations (**Addition**, **Division**, **Multiplication** and **Subtraction**) can be pictured as a family. **Mr Addition** been the man realised that he is not good at letting go of things so he decided to marry a woman who has the ability to let go as her primary area of strength. Well, **Mr Addition** marries **Miss Subtraction** and they both decided to have two children. In no time, **Mrs Subtraction** popped the good news to her darling husband, *I'm pregnant. Yippee!* **Mr Addition** jubilated round the house, he was so excited that he will welcome the *little him* couple of months from then.

Mr **Addition** wished their coming child will be a prototype of him and **Mrs Subtraction** wished so too, so one day while chatting both of them expressed their desire of the coming baby as 50% of Dad and 50% of Mum.

Days turned into weeks and weeks into months, **Mrs Subtraction** laid on the delivery bed and was told, *Madam you need to push*. She pushed and a little cutie was born, the couple were more than excited to welcome a new member to the family.

The baby was a carbon copy of **Mr Addition** but in a more intense way, hence, they christened the baby **Multiplication**. **Mrs Subtraction**, though excited, still desired a child that will look more like her, so when she got pregnant, this time, the baby looked like her so much, hence, the baby was christened **Division**.

Yeah!!! The couple is very happy; each of them got their prototype in their children.

What are we trying to establish?

Multiplication is repeated Addition while Division is repeated Subtraction. Firstly, let us consider Multiplication as repeated Addition with the aid of examples.

1   $2 \times 3 = 6$ means;

    a)   Add 2 three times

        $2 + 2 + 2 = 6$

or

b) Add 3 two times

3 + 3 = 6

2 3 × 5 = 15 means;

a) Add 3 five times

3 + 3 + 3 + 3 + 3 = 15

or

b) Add 5 three times

5 + 5 + 5 = 15

3 4 × 8 = 32 means;

a) Add 4 eight times 4 + 4 + 4 + 4 + 4 + 4 + 4 + 4 = 32

or

b) Add 8 four times 8 + 8 + 8 + 8 = 32

When multiplying by 0, the same process applies. All you need to do is interpret 0 to be *nothing.*

1 7 × 0 = 0 means;

a) Add 7 zero times, that is do not add 7 at all

or

b) Add 0 seven times 0 + 0 + 0 + 0 + 0 + 0 = 0

2 10 × 0 = 0 means;

a) Add 10 zero times that is do not add 10 at all

b) Add 0 ten times 0 + 0 + 0 + 0 + 0 + 0 + 0 + 0 + 0 + 0 = 0

Multiplication is commutative in nature.
For instance,
2 × 4 = 8
4 × 2 = 8
Multiplication is repeated addition.
For instance,
2 × 4 = 8 means 2 + 2 + 2 + 2 = 8
or
4 + 4 = 8

Remember the story says Multiplication took after Mr Addition while Division looks like Mrs Subtraction, hence, it is time to talk about Division being a copycat of Mrs Subtraction.

Just like Multiplication is repeated Addition, Division is also repeated Subtraction.

How do you go about it?

Keep subtracting the number (or denominator) from the dividend (or numerator) till you get 0 or till you can no longer subtract the divisor anymore.

Examples are great ways of making Mathematics real, let's consider some.

1    $6 \div 3$ means;

    Subtract 3 from 6 till you get 0

    6-3=3

    3-3=0

    Or

    6-3-3=0

    How many times was 3 subtracted from 6 to get 0? 2

    $6 \div 3 = 2$

2    $15 \div 5$ means;

    Subtract 5 from 15 till you get 0

    15-5=10

    10-5=5

    5-5=0

    Or

    15-5-5-5=0

    How many times was 5 subtracted from 15 to get 0? 3

    $15 \div 5 = 3$

3    $84 \div 6$ means;

    Subtract 6 from 84 till you get 0

    84-6=78   1

    78-6=72   2

    72-6=66   3

    66-6=60   4

    60-6=54   5

    54-6=48   6

    48-6=42   7

42-6=36   8
36-6=30   9
30-6=24   10
24-6=18   11
18-6=12   12
12-6=6   13
6-6=0   14
Or

84-6-6-6-6-6-6-6-6-6-6-6-6-6-6=0

How many times was 6 subtracted from 84 to get 0? 14

$84 \div 6 = 14$

What happens when the divisor does not divide the dividend perfectly? Let's consider some examples:

1. $25 \div 4$ means;

   25-4=21   1
   21-4=17   2
   17-4=13   3
   13-4=9   4
   9-4=5   5
   5-4=1   6

   5 cannot be subtracted from 1

   $25 \div 4 = 6$ remainder 1

2. $33 \div 7$ means;

   33-7= 26   1
   26-7=19   2
   19-7=12   3
   12-7=5   4

   7 cannot be subtracted from 5

   $33 \div 7 = 4$ remainder 5

Introduction

It is important to remember Place-value System and Addition in order to fully understand multiplication.

The value of anything including numbers lies in its worth and the worth of any digit is determined by where it is placed or positioned in a number.

Every number is made up of the ten basic numbers which are 0,1,2,3,4,5,6,7,8,9 and all of which are significant including 0.

Zero can make a huge difference monetarily; take for instance you have ₦1,000 and 0 is placed behind it, it becomes ₦10,000 if another 0 is placed behind it, it becomes ₦100,000. This indicates that **no number is insignificant**. Addition on the other hand, has to do with counting numbers together.

Units

This is often represented by **U**.

The following define Units;
1. The first digit from the right in any number e.g. 23**4**.
2. Lowest positive natural number
3. A set containing a single number e.g. **3**.
4. The first digit to the left of the decimal point in decimal notation, representing a whole number less than ten e.g. 73**8**.25.

Tens

Tens is often represented by **T**.

The following define Tens;
1. The second digit from the right in any number e.g. 2**3**4.
2. The second digit to the left of the decimal point in decimal notation e.g. 7**3**8.25.

Hundreds

Hundreds is often represented by **H**.

The following define Hundreds;
1. The third digit from the right in any number e.g. **2**34.
2. The third digit to the left of the decimal point in decimal notation e.g. **7**38.25.

Thousands

Thousands is often represented by the combination of **T** and **h**, that is, **Th**.

The following define Thousands;
1. The fourth digit from the right in any number e.g. 16 9**0**2 831
2. The fourth digit to the left of the decimal point in decimal notation e.g. 21**4** 738.25.

3   A comma is sometimes placed after every three digits from the right, hence, the digit after the first comma is Thousands e.g. **1**, 093.

Tens of Thousands

Tens of Thousands is often represented by two **T** and **h**, **TTh**; which is a combination of Tens **T** and Thousands **Th**.

The following defines Tens of Thousands;
1   The fifth digit from the right in any number e.g. 16 **9**02 831.
2   The fifth digit to the left of the decimal point in decimal notation e.g. **2**14738.25

Hundreds of Thousands

Hundreds of Thousands is often represented by **H**, **T** and a small letter **h**, **HTh**; which is a combination of Hundreds "H" and Thousands **Th**.

The following defines Hundreds of Thousands;
1   The sixth digit from the right in any number e.g. 16 **9**02 831
2   The sixth digit to the left of the decimal point in decimal notation e.g. **2**14738.25.

Millions

Millions is often represented by **M**.

The following defines Millions;
1   The seventh digit from the right in any number e.g. 16 902 831.
2   The seventh digit to the left of the decimal point in decimal notation e.g. **4** 202 387.25.
3   The digit after the second comma from the right e.g. **3** 359 201.

Tenths

Tenths is often represented by **t**.

Tenths is the digit immediately after the decimal point to the right hand side (RHS). E.g. 2 387.**2**5.

Hundredths

Hundredths is often represented by **h**.
Hundredths is the second digit to the right hand side of the decimal point. E.g. 2387.2**5**

Partitioning

Another way to grasp the place-value system is to classify numbers into sections. These sections are informed by the points where commas are placed.

Billion Section			Million Section			Thousand Section			Units Section		
Hundreds	Tens	Units	Hundreds	Tens	Units	Hundreds	Tens	Units	Hundreds	Tens	Units

Example 1.1

Identify the values of the circled digits in the numbers below;

(a) 2,3①9

The value of 1 is Tens.

(b) 9⓪1,788

0-ten thousands

(c) 45.2③8

3-hundredths

(d) 1,⓪00,000

0-hundreds of thousand

Example 1.2

Write the values of all the digits in the numbers given below;

(a) 290,189

	HTh	TTh	Th	H	T	U
9 Units = 9						9
8 Tens = 80					8	0
1 Hundreds = 100				1	0	0
0 Thousands = 0000			0	0	0	0
9 Ten Thousands = 90,0009		9	0	0	0	0
2 Hundreds of Thousands = 200,0002	2	0	0	0	0	0

(b) 7,126,345.923

	M	HTh	TTh	Th	H	T	U	t	h	th
3 Thousandths = 0.003										3
2 Hundredths = 0.02									2	0
9 Tenths = 0.9								9	0	0
5 Units = 5							5	0	0	0
4 Tens = 40						4	0	0	0	0
3 Hundreds = 300					3	0	0	0	0	0
6 Thousands = 6,000				6	0	0	0	0	0	0
2 Ten Thousands = 20,000			2	0	0	0	0	0	0	0
1 Hundreds of Thousands =100,000		1	0	0	0	0	0	0	0	0
7 Millions= 7,000,00	7	0	0	0	0	0	0	0	0	0

Do It Yourself 1

1 What is the value of 9 in the following numbers?
   (a) 67,690
   (b) 3452.39
   (c) 75,389,634
   (d) 6,895,564
   (e) 956,657,887

2 Write the values of all the digits in the following numbers;
   (a) 12,162
   (b) 76,133,543.611

3 State the values of the digits that occur more than once in these numbers
   (a) 8980.12
   (b) 92,390.22

As earlier stated, multiplication is repeated addition hence there is a need to refresh the understanding of addition especially those that involve 'carrying over'.

Example 2.1

Add 64 and 89.

H	T	U
	6	4
+	8	9
		3

4 + 9 = 13, write 3 and keep 1

H	T	U
	6	4
+	8	9
1	5	3

6 + 8 = 14, 14 + 1 (earlier kept) = 15

Example 2.2

Add 7843 and 1208.

Th	H	T	U
7	8	4	3
+ 1	2	0	8
			1

3 + 8 = 11; write 1 and keep 1

Th	H	T	U
7	8	4	3
+ 1	2	0	8
		5	1

4 + 0 = 4, 4 + 1 (earlier kept) = 5

	Th	H	T	U	
	7	8	4	3	
+	1	2	0	8	
		0	5	1	8 + 2 = 10; write 0 and keep 1

	Th	H	T	U	
	7	8	4	3	
+	1	2	0	8	
	9	0	5	1	7 + 1 = 8, 1 (earlier kept) + 8 = 9

Multiplication techniques include;

1 Body part multiplication

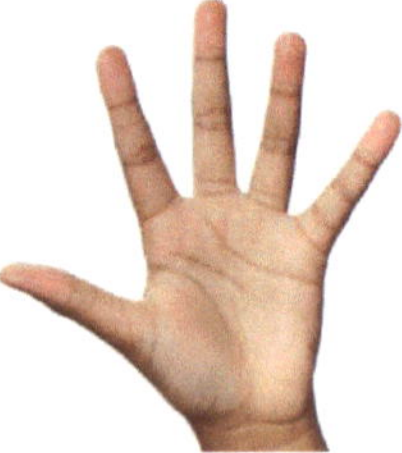

2 Line or graph multiplication

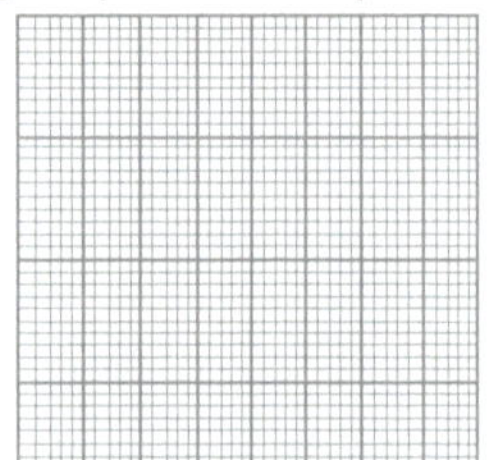

3 Diamond multiplication

4 Box multiplication

5 Mental multiplication

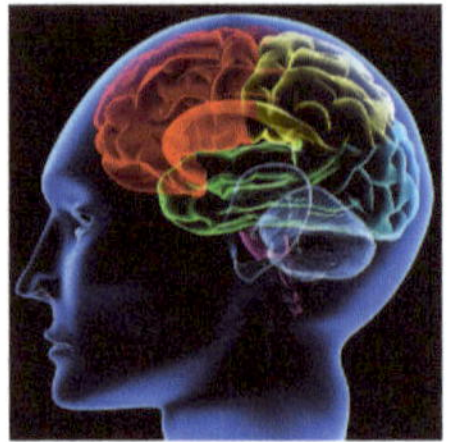

Chapter 3 Body part Multiplication

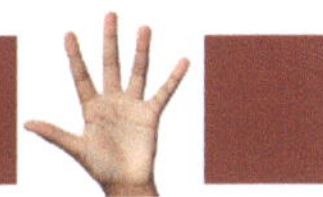

It is most exciting to know that God equipped us with calculator when he created us, however, many are not aware of this. This book will serve as an eye opener to this in-built calculator.

6 Times Table

This entails the use of all fingers. However, it is only applicable for 6×6 to 6×10. Place your hands horizontally such that the tips of both middle fingers are almost touching each other.

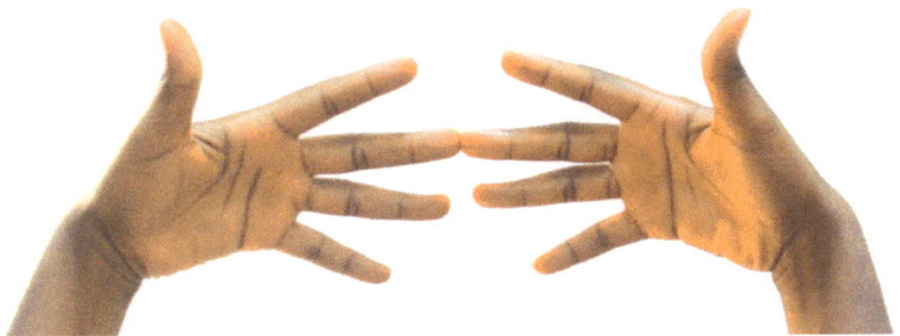

Step 1: Label all fingers 6 to 10 in this format;
Pinkie - 6
Ring finger - 7
Middle finger - 8
Index finger - 9
Thumb -10

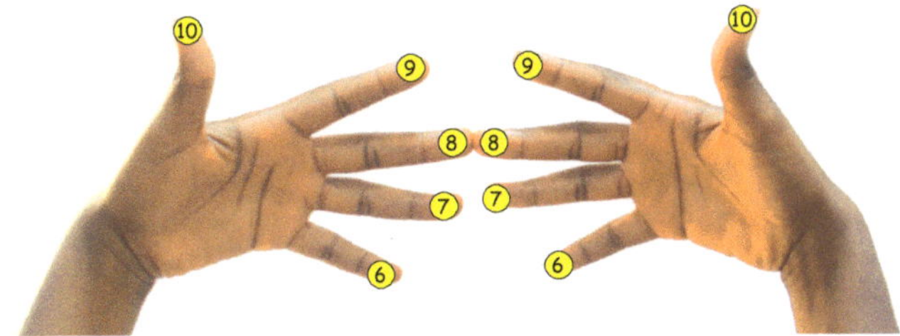

Step 2: Let the two fingers to be multiplied touch each other.

Step 3: Count the fingers above the two joined together and multiply the value on the right hand by the value on the left hand. This will represent the Units/Ones.

Step 4: Count every other number not multiplied above which includes the fingers joined together. This will represent the Tens.

Step 5: Add both numbers to obtain the answer.

Simplify 6 × 6.

Fingers touch each other.

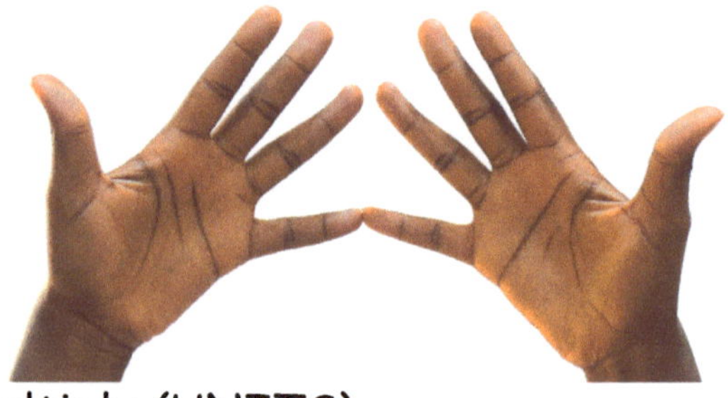

Count and multiply (UNITS)

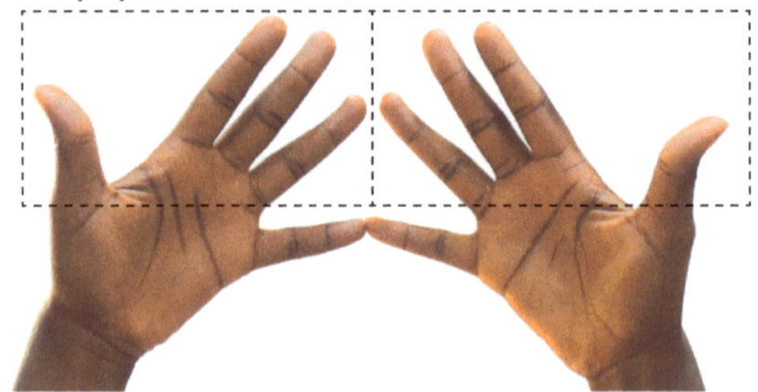

4 × 4 = 16

Count (TENS)

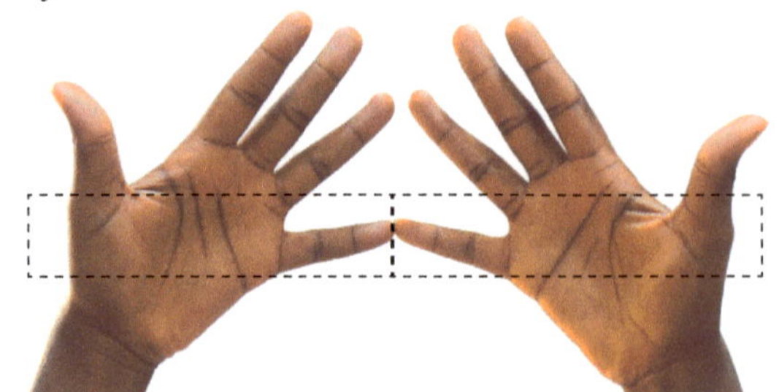

1 + 1 = 2

Add both numbers to obtain the answer.

$$
\begin{array}{r r}
\text{T} & \text{U} \\
2 & \\
+ \quad 1 & 6 \\
\hline
3 & 6 \\
\hline
\end{array}
$$

6 × 6 = 36

Evaluate 6 × 7.

Fingers touch each other.

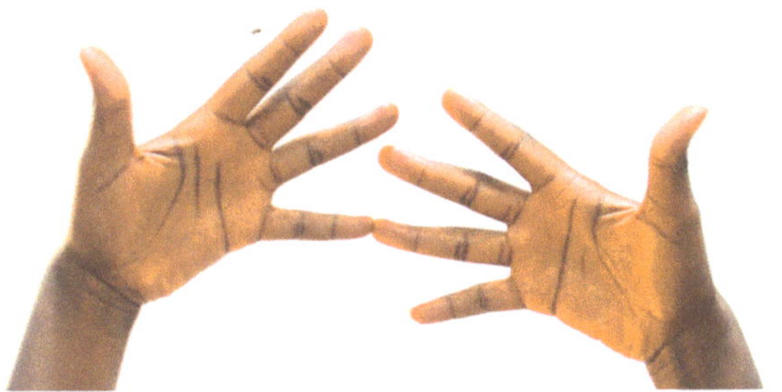

Count and multiply (UNITS)

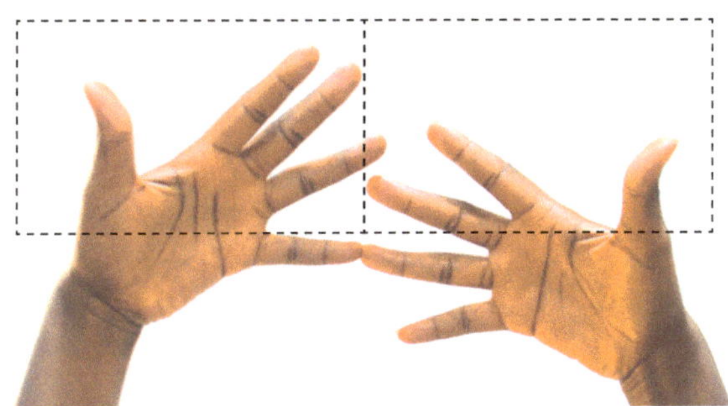

4 × 3 = 12

Count (TENS)

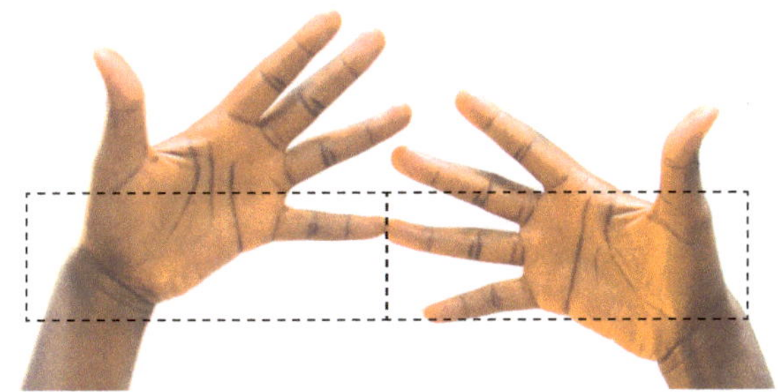

1 + 2 = 3

Add both numbers to obtain the answer.

	T	U
	3	
+	1	2
	4	2

6 × 7 = 42

Multiply 6 by 9.

Fingers touch each other.

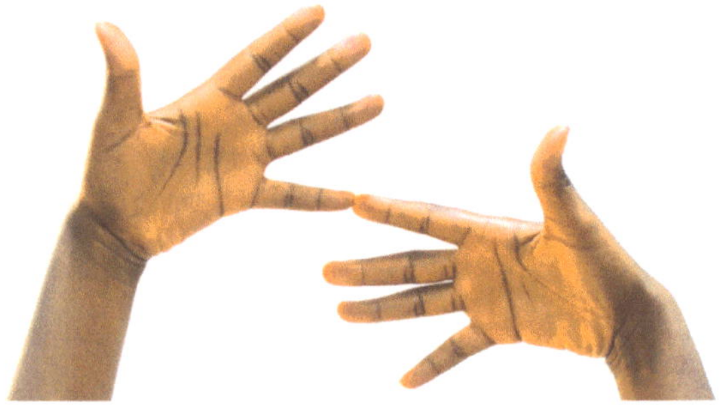

Count and multiply (UNITS)

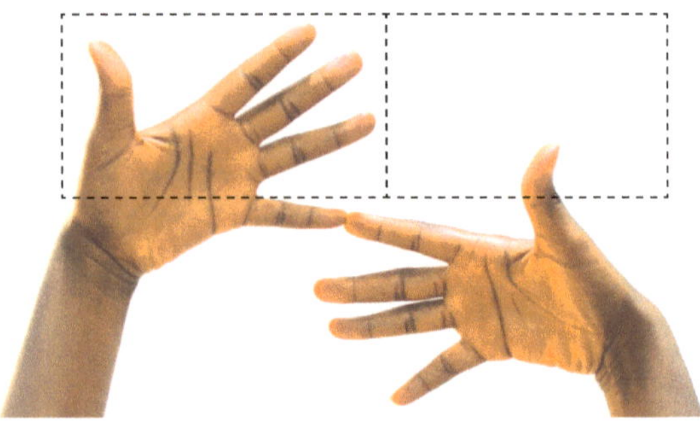

$4 \times 1 = 4$

Count (TENS)

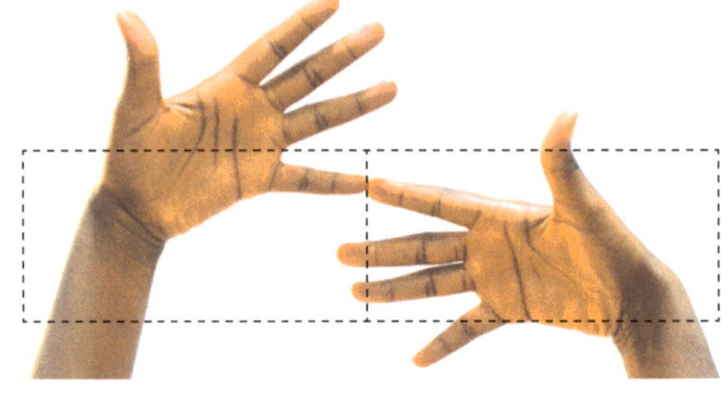

$1 + 4 = 5$

Add both numbers to obtain the answer

T	U
5	
+	4
5	4

$6 \times 9 = 54$

7 Times Table

This entails the use of all fingers. However, it is only applicable for 7 × 6 to 7 × 10. Place your hands horizontally such that the tips of both middle fingers are almost touching each other

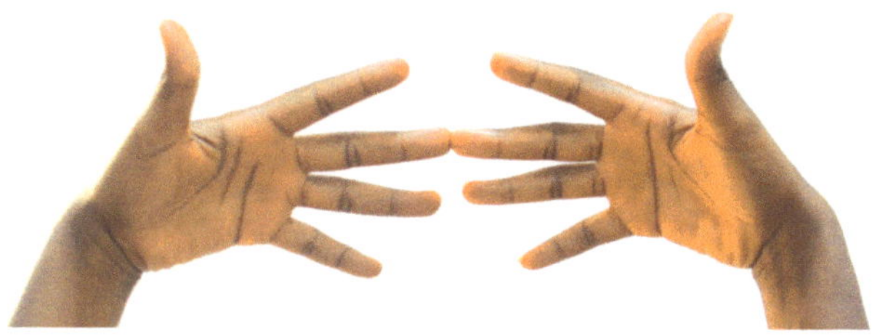

Step 1: Label all fingers 6 to 10 in this format;
Pinkie -6
Ring finger-7
Middle finger-8
Index finger-9
Thumb -10

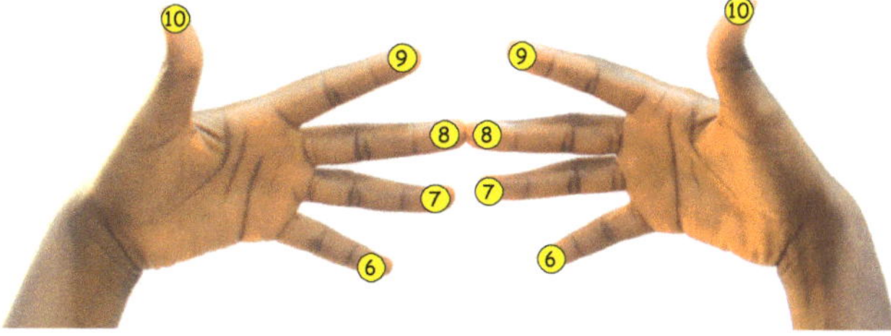

Step 2: Let the two fingers to be multiplied touch each other.

Step 3: Count the fingers above the two joined together and multiply the value on the right hand by the value on the left hand. This will represent the Units/Ones.

Step 4: Count every other number not multiplied above which includes the fingers joined together. This will represent the Tens.

Step 5: Place both numbers beside each other to obtain the answer.

Solve 7 × 6.

Fingers touch each other.

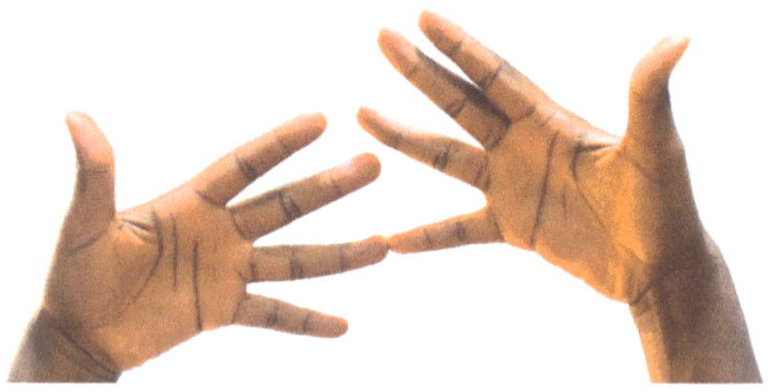

Count and multiply (UNITS)

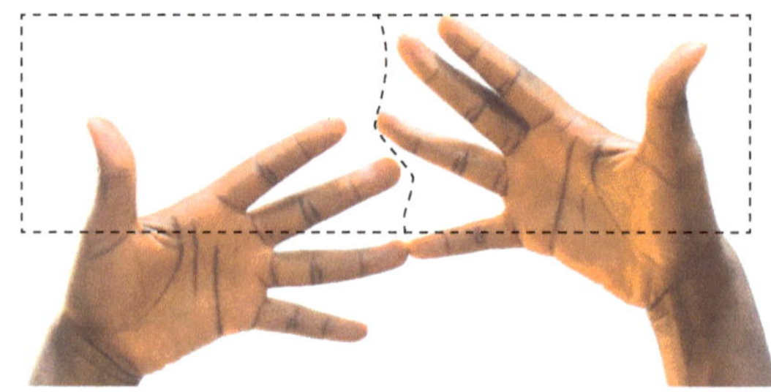

3 × 4 = 12

Count (TENS)

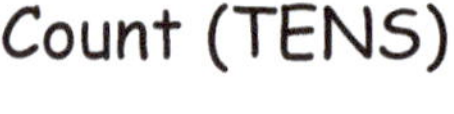

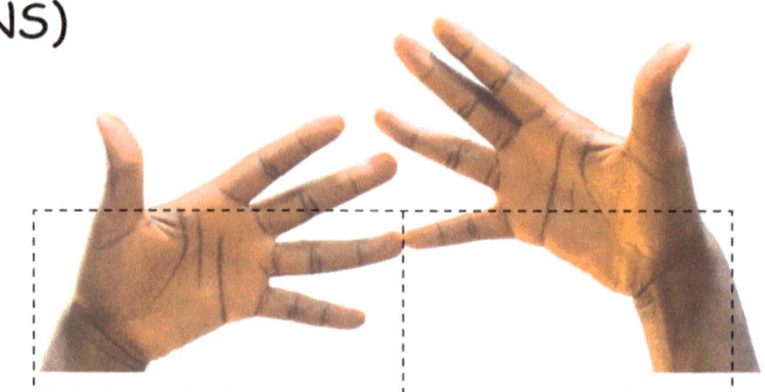

2 + 1 = 3

Add both numbers to obtain the answer

```
      T      U
      3
  +   1      2
  ___________
      4      2
  ___________
```

7 × 6 = 42

You will observe that this is the same as 6 × 7 which verifies the **commutative attribute** of multiplication.

Solve 7 × 8.

Fingers touch each other.

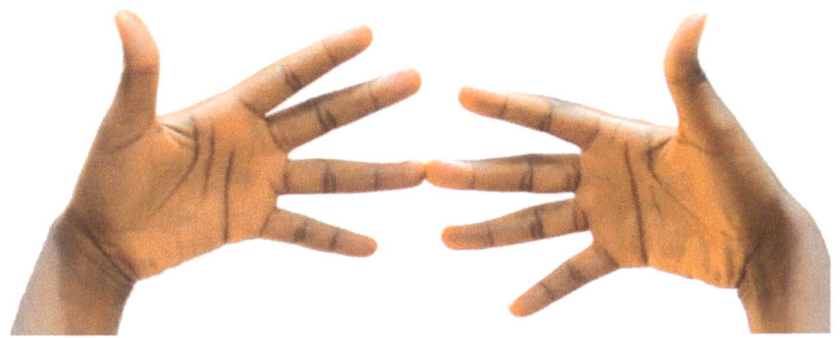

Count and multiply (UNITS)

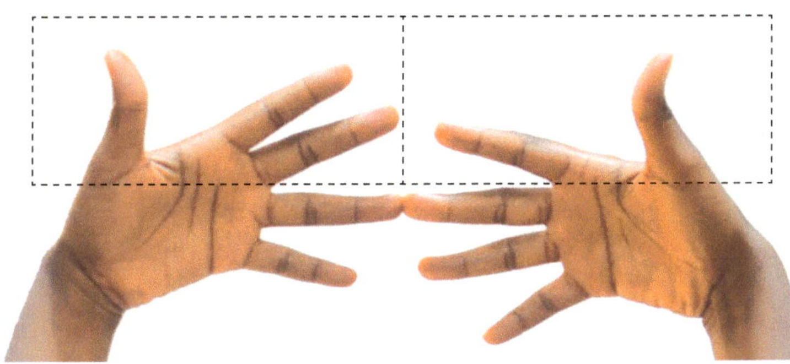

3 × 2 = 6

Count (TENS)

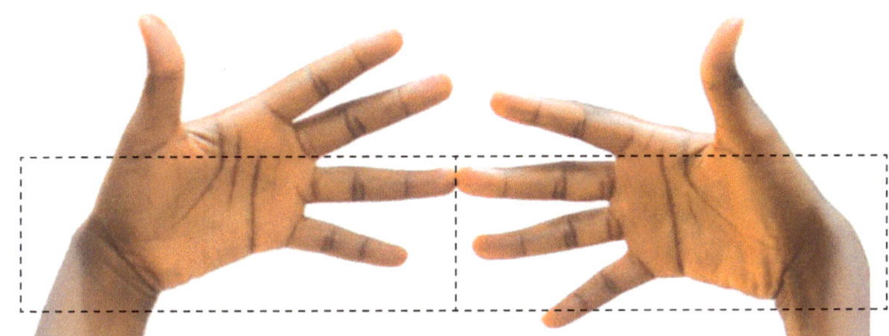

2 + 3 = 5

Add both numbers to obtain the answer

	T	U
	5	
+		6
	5	6

7 × 8 = 56

Solve 7 × 10.

Fingers touch each other.

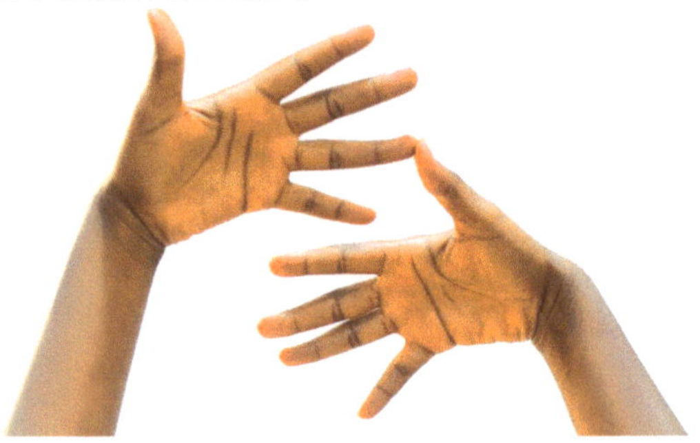

Count and multiply (UNITS)

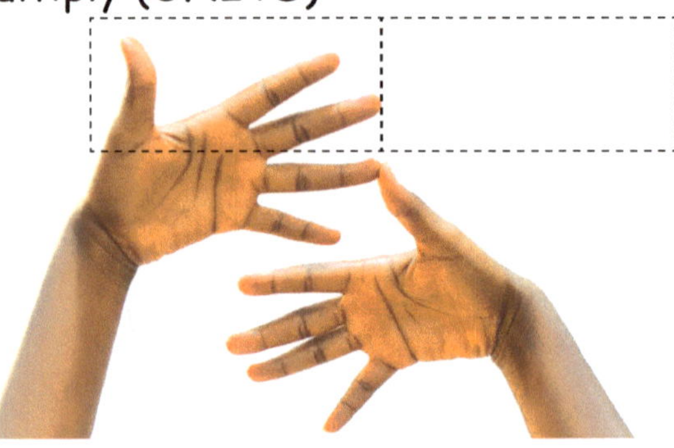

3 × 0 = 0

Count (TENS)

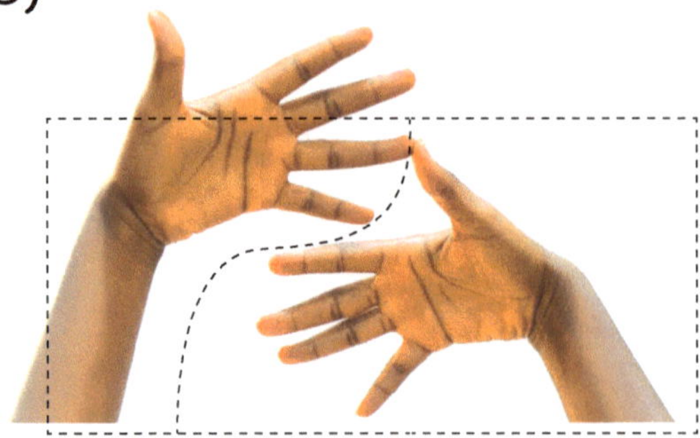

2 + 5 = 7

Place both numbers beside eachother to obtain the answer

7 × 10 = 70

Do It Yourself 3

Multiply the following with your fingers.

1   7 × 7
2   7 × 9

8 Times Table

8 × 1 to 8 × 5

Step 1: Open your palms and label only the left hand 1 to 5.

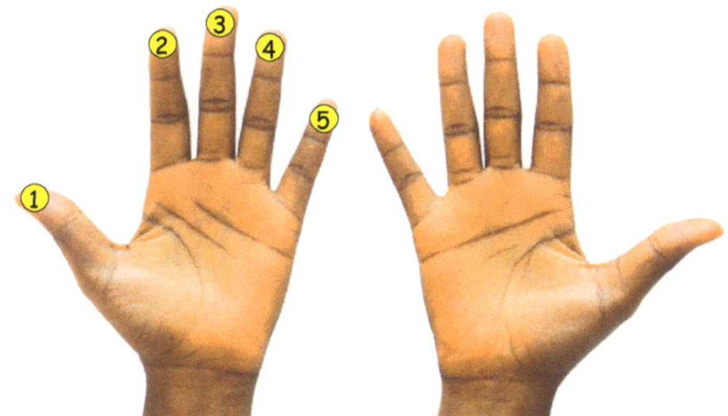

Step 2: Bend the finger you are multiplying 8 with on the left and bend the number of the finger on the right.

Step 3: Count the unbent fingers in groups

Example 3.7

Solve 8 × 1.

Bend the finger labelled "1" on the left and bend one finger on the extreme right.

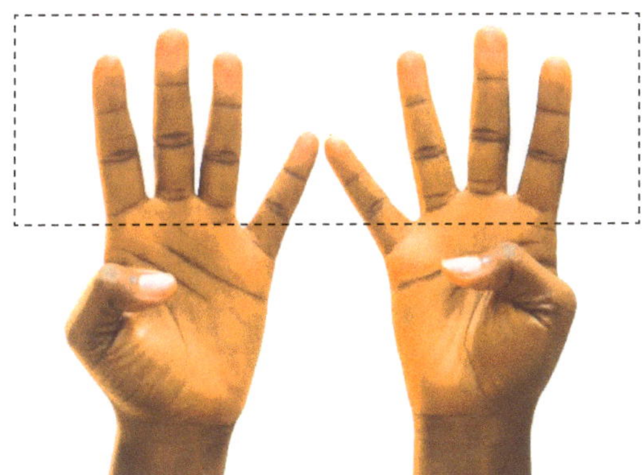

Count the unbent fingers.

8 × 1 = 8

Solve 8 × 2.

Bend the finger labelled "2" on the left and bend two fingers on the extreme right.

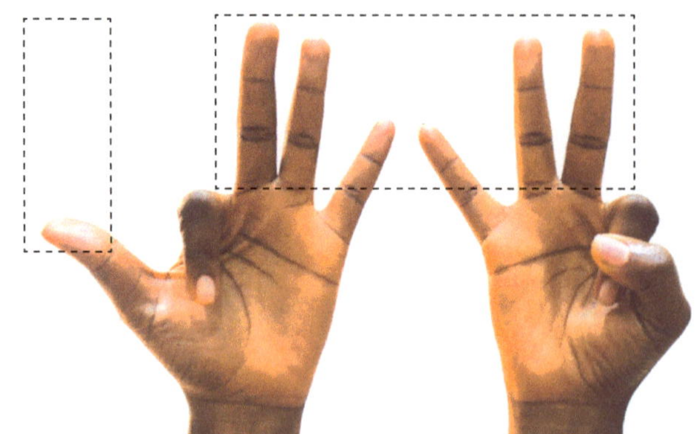

Count the unbent fingers.
8 × 2 = 16

Example 3.9

Solve 8 × 4.

Bend the finger labelled "4" on the left and bend four fingers on the extreme right.

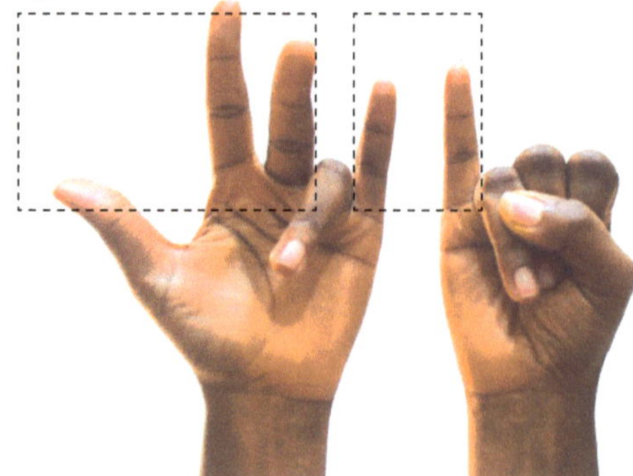

Count the unbent fingers
8 × 4 = 32

Do It Yourself 4

Do the following.
1   8 × 3
2   8 × 5

8 × 6 to 8 × 10

This entails the use of all fingers. However, it is only applicable for 8 × 6 to 8 × 10. Place your hands horizontally such that the tips of both middle fingers are almost touching each other.

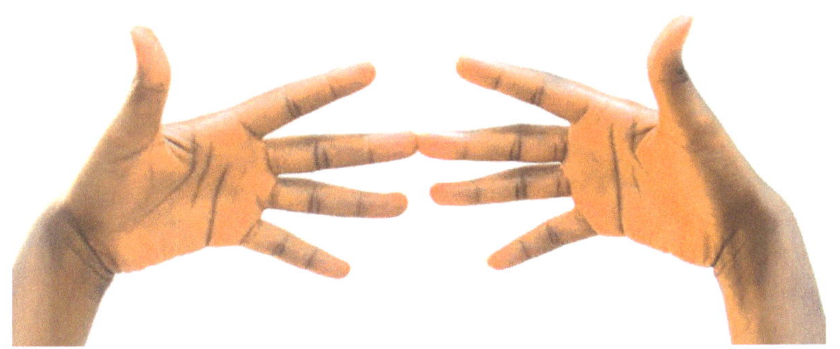

Step 1: Label all fingers 6 to 10 in this format;
Pinkie -6
Ring finger-7
Middle finger-8
Index finger-9
Thumb -10

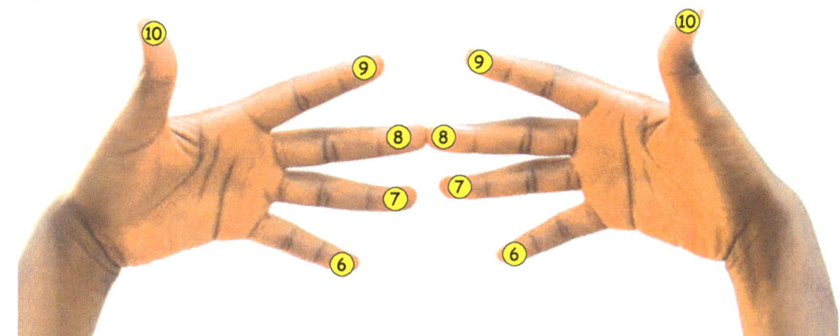

Step 2: Let the two fingers to be multiplied touch each other.

Step 3: Count the fingers above the two joined together and multiply the value on the right hand by the value on the left hand. This will represent the Units/Ones.

Step 4: Count every other number not multiplied above which includes the fingers joined together. This will represent the Tens.

Step 5: Place both numbers beside eachother to obtain the answer.

Solve 8 × 6.

Fingers touch each other.

Count and multiply (UNITS)

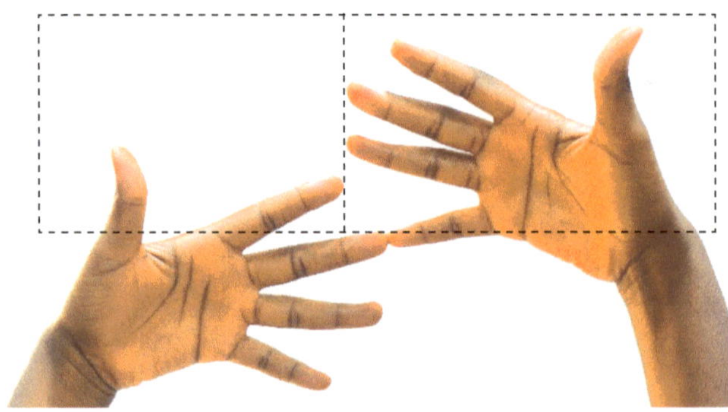

2 × 4 = 8

Count (TENS)

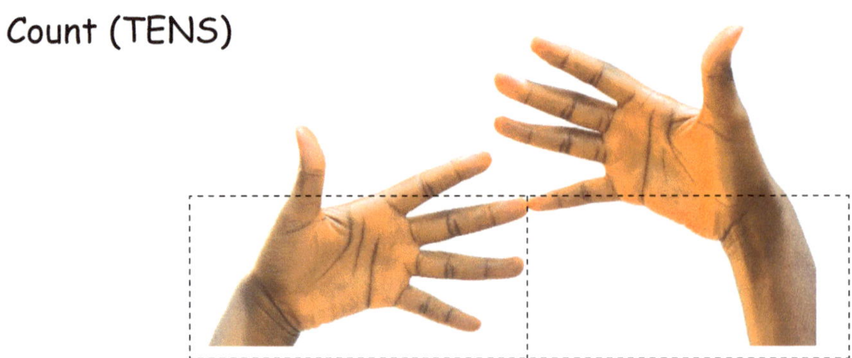

3 + 1 = 4

Place both numbers beside each other to obtain the answer.
8 × 6 = 48

Solve 8 × 7.

Fingers touch each other.

Count and multiply (UNITS)

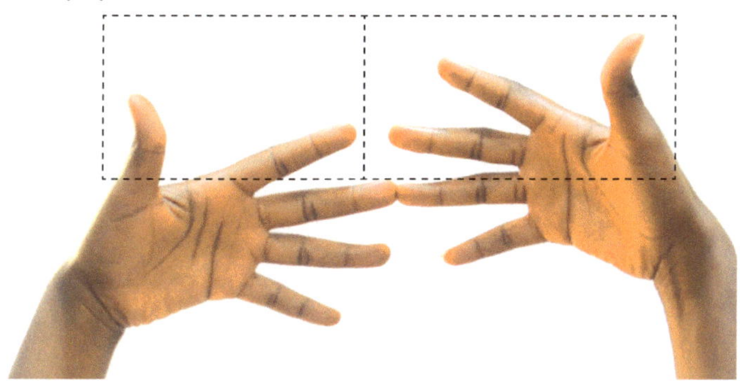

2 × 3 = 6

Count (TENS)

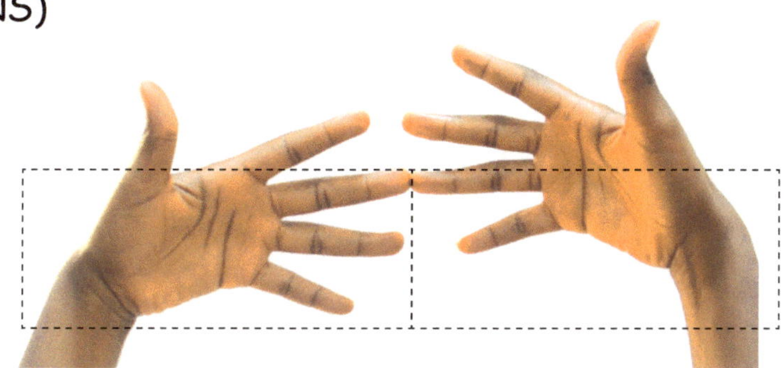

3 × 2 = 6

Place both numbers beside each other to obtain the answer.
8 × 7 = 56

Solve 8 × 8.

Fingers touch each other.

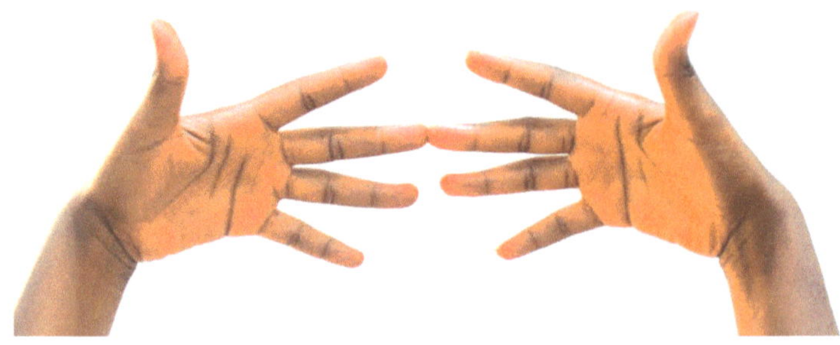

Count and multiply (UNITS)

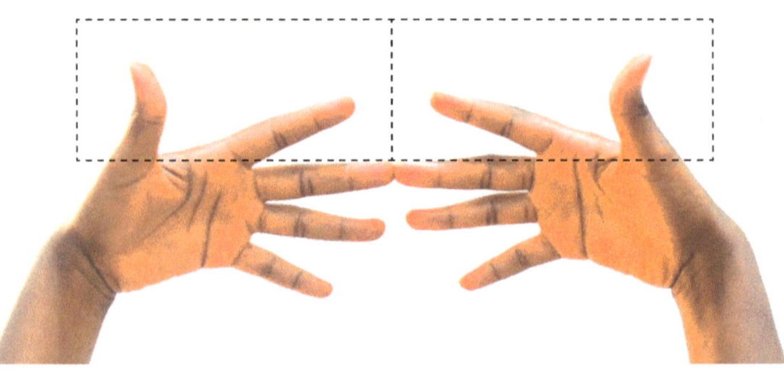

2 × 2 = 4

Count (TENS)

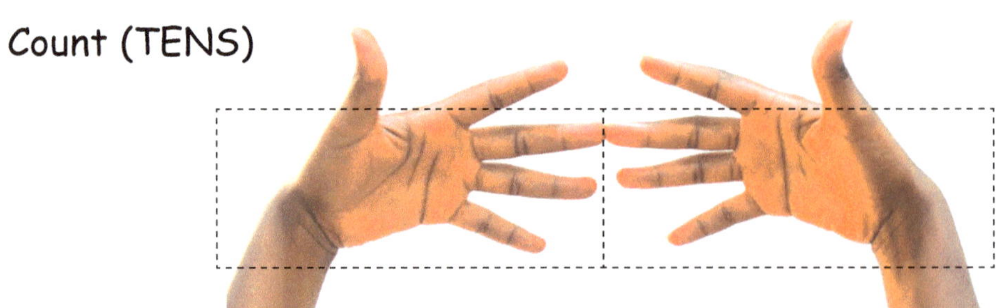

3 + 3 = 6

Place both numbers beside each other to obtain the answer.

8 × 8 = 64

Do It Yourself 5

Using natural calculator, carry out the multiplications below;

1 8 × 9
2 8 × 10

9 Times Table

Finger Multiplication

Step 1:   Open your palms such that it is facing or backing you.

Step 2:   Label your fingers 1 to 10 in this order;
If your palms are facing you:
Left thumb - 1
Left index finger - 2
Left middle finger - 3
Left ring finger - 4
Left pinkie - 5
Right pinkie - 6
Right ring finger - 7
Right middle finger - 8
Right index finger - 9
Right thumb - 10

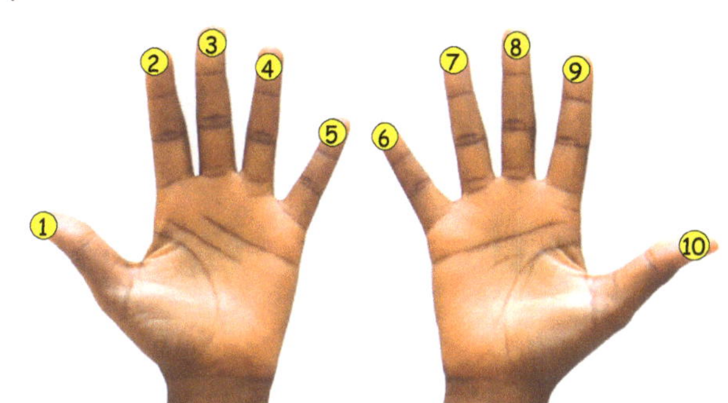

If your palms are backing you:
Left pinkie - 1
Left ring finger - 2
Left middle finger - 3
Left index finger - 4
Left thumb - 5
Right thumb - 6
Right index finger - 7
Right middle finger - 8
Right ring finger - 9
Right pinkie - 10

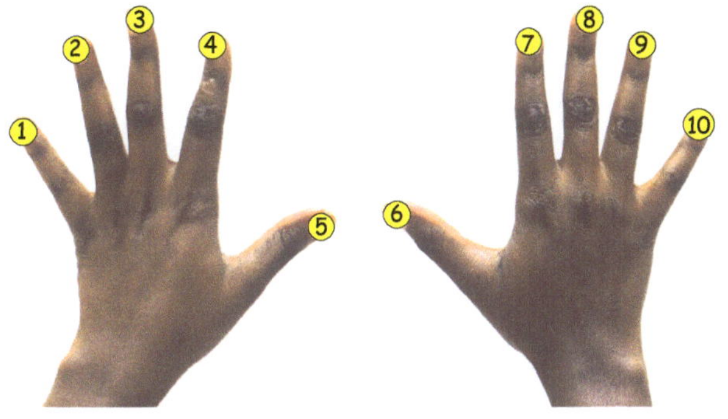

NOTE: In this book, we shall make use of the first labelling. However, if the same principle is applied to the second labelling, it will yield exactly the same result.

Step 3: Bend the finger multiplied with 9. For instance, if you are multiplying 9 × 3, bend the finger labelled 3. Please note that you are not to bend three fingers, rather the finger labelled three, that is, the middle finger on your left hand.

Step 4: Count the number of fingers on the Left Hand Side (LHS) of the bent finger which will be the Tens of the answer and the Right Hand Side (RHS) of the bent finger which will be the Units/Ones of the answer. Please note that Left Hand Side (LHS) is not left hand and Right Hand Side (RHS) is not right hand.

Example 3.13
Solve 9 × 1.
Bend the finger labelled 1.

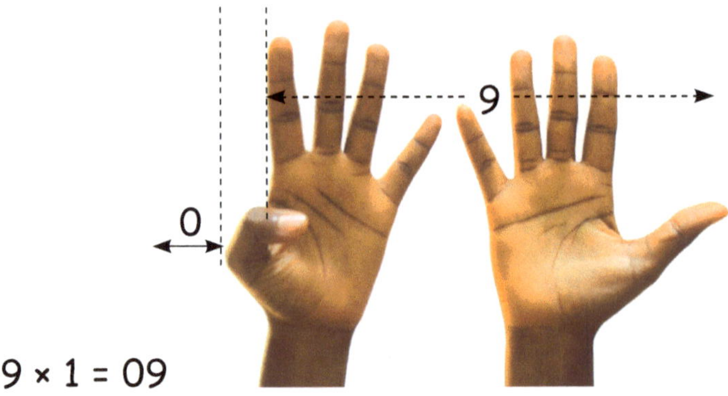

9 × 1 = 09

Example 3.14
Solve 9 × 2.
Bend the finger labelled 2.

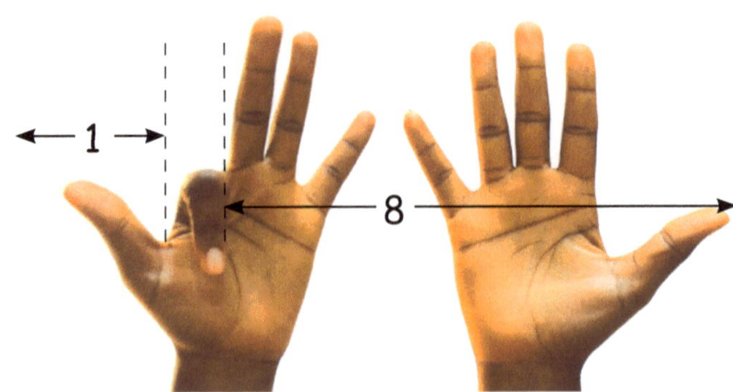

9 × 2 = 18

Example 3.15
Solve 9 × 4.
Bend the finger labelled 4

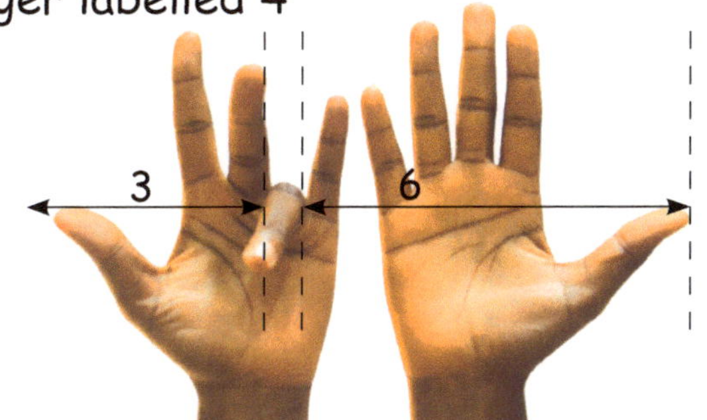

9 × 4 = 36

Example 3.16
Solve 9 × 8.

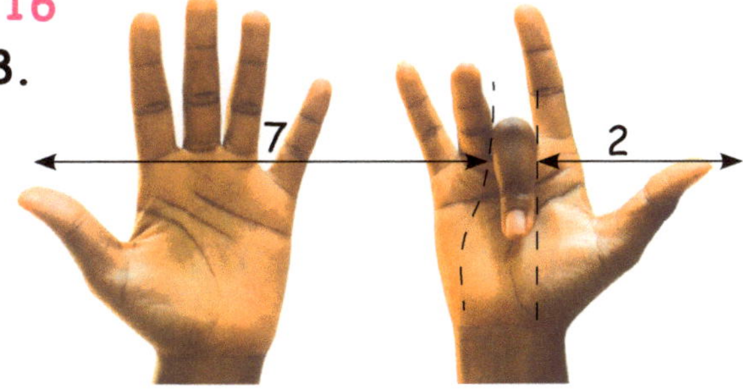

Bend the finger labelled 8.
9 × 8 = 72

Example 3.17
Solve 9 × 10.
Bend the finger labelled 10.

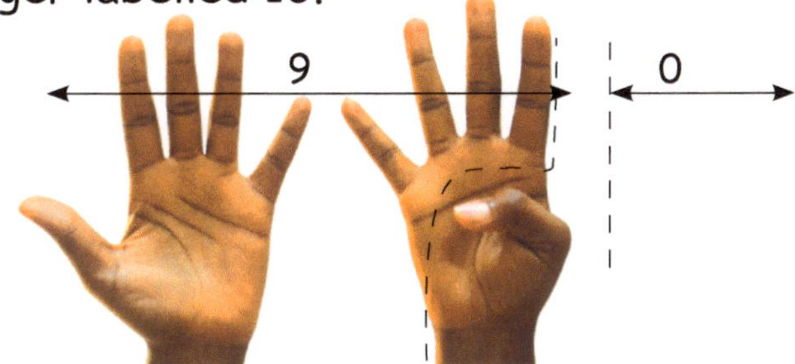

9 × 10 = 90

Do It Yourself 6

Use your fingers to multiply the following;

1 9 × 3   **2** 9 × 5   **3** 9 × 6   **4** 9 × 7   **5** 9 × 9

Finger and Toe Multiplication

So far, we have only used fingers to multiply, now we are including the toes to enable us multiply higher numbers.

Step 1:   Open your palms and mentally place it beside your toes.

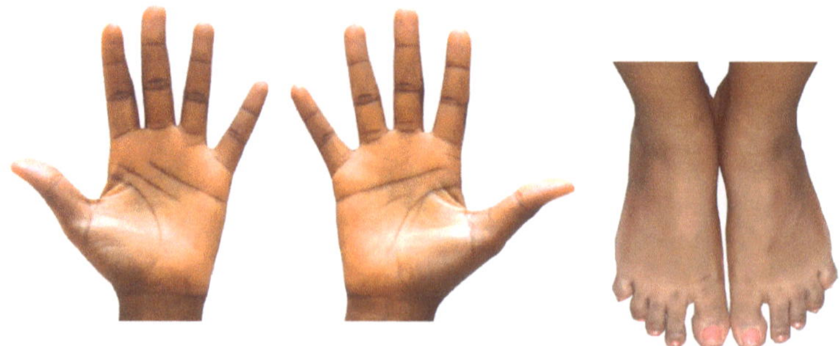

Step 2:   Label your fingers 1 to 10 just as you did above and label your toes 1 to 10 the way they are placed on the floor. The fingers represent Tens while the toes represent Units/Ones.
For instance, in 9 × 14, 1 will be the finger labelled 1 while 4 will be the toe labelled 4.

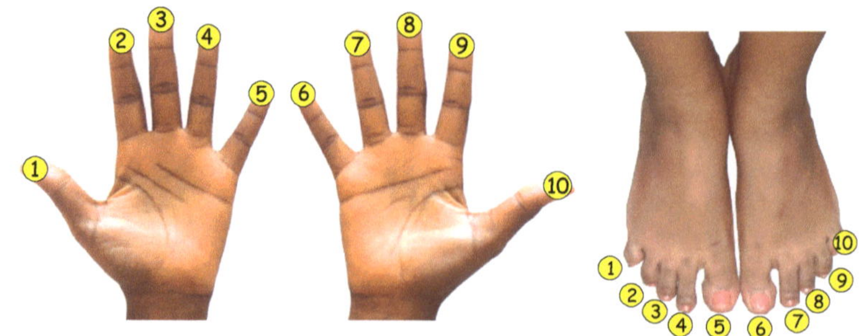

Step 3:   Bend the finger and toe (or mentally shade) under consideration.

Step 4:   Count the finger(s) and toe(s) between the bent finger and shaded toe, this will represent Tens. Count the remaining toes after the shaded toe, this will be the Units/Ones.

Example 3.18

Solve 9 × 11.

Bend the finger labelled 1 and shade the toe labelled 1.

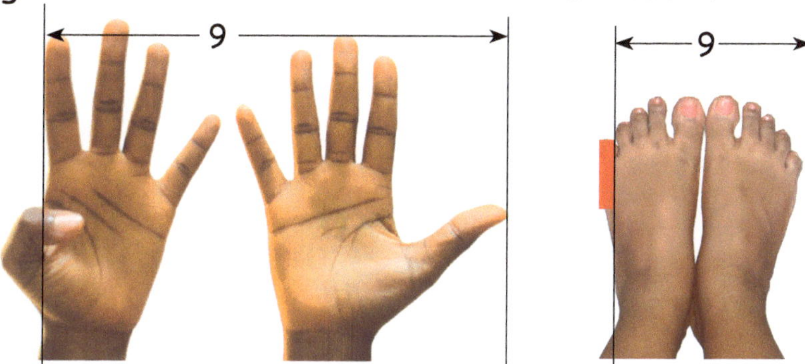

9 × 11 = 99

Example 3.19

Solve 9 × 12.

Bend the finger labelled 1 and shade the toe labelled 2.

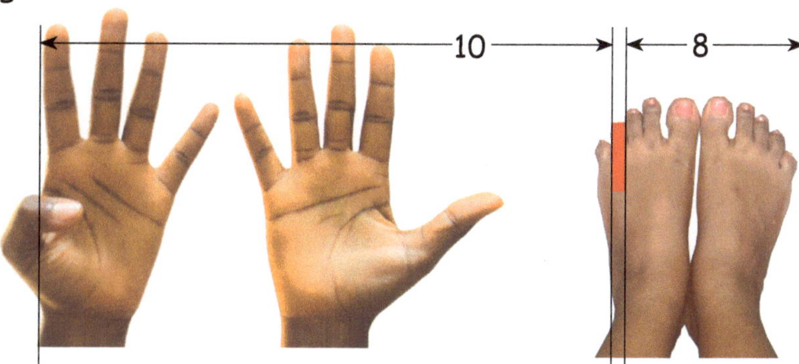

9 × 12 = 108

Example 3.20

Solve 9 × 16.

Bend the finger labelled 1 and shade the toe labelled 6.

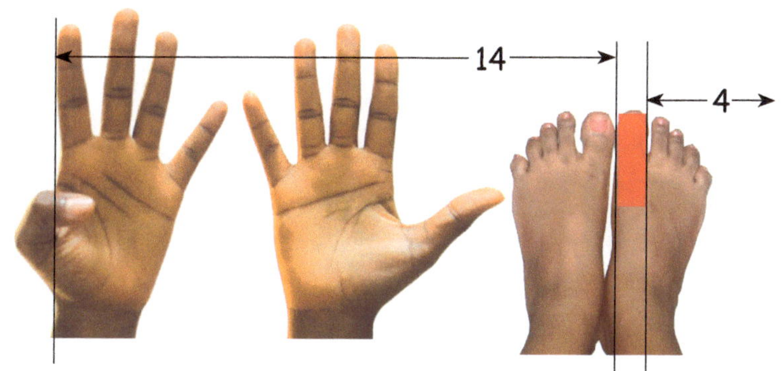

9 × 16 = 144

Example 3.21

Solve 9 × 20.

Bend the finger labelled 2 and do not shade any toe. This is slightly different because no toe is shaded.

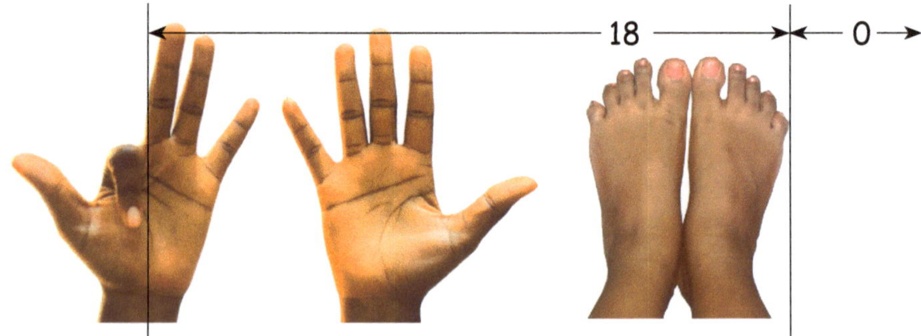

9 × 20 = 180

Do It Yourself 7

1	9 × 13		**2**	9 × 14		**3**	9 × 15
4	9 × 17		**5**	9 × 18		**6**	9 × 19

10 Times Table

This entails the use of all fingers. However, it is only applicable for 10 × 6 to 10 × 10. To do this, bend your hands such that the tips of the middle finger are almost touching each other.

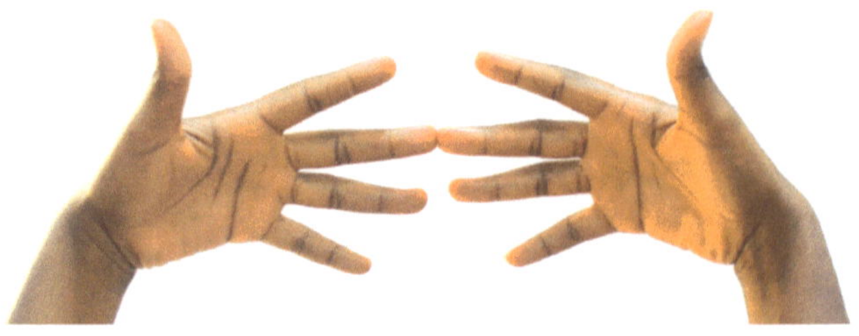

Step 1: Label all fingers 6 to 10 in this format;
Pinkie - 6
Ring finger - 7
Middle finger - 8
Index finger - 9
Thumb - 10

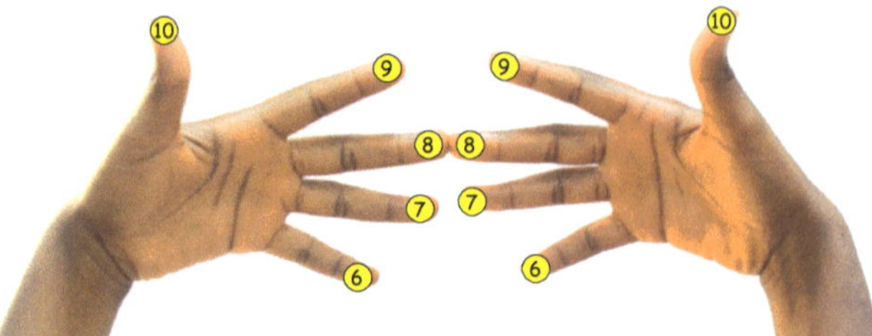

Step 2: Let the two fingers to be multiplied touch each other.

Step 3: Count the fingers above the two joined together and multiply the value on the right hand by the value on the left hand. This will represent the Units/Ones.

Step 4: Count every other number not multiplied above which includes the fingers joined together. This will represent the Tens.

Step 5: Place both numbers beside eachother to obtain the answer

Example 3.22

Solve 10 × 6.

Fingers touch each other.

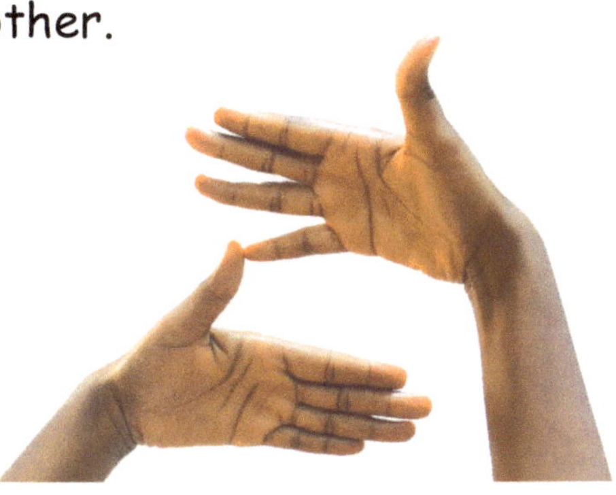

Count and multiply (UNITS)

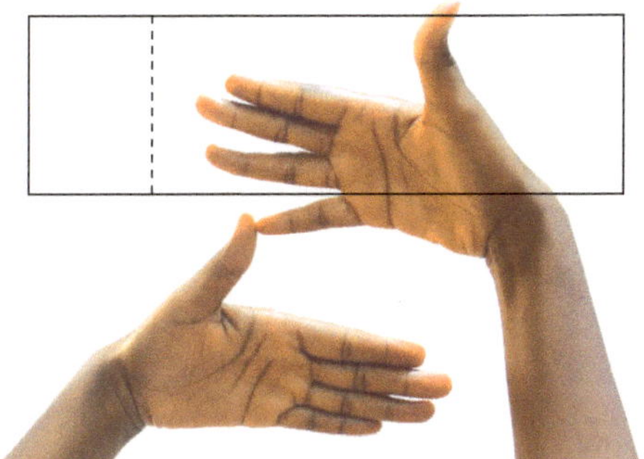

0 × 4 = 0

Count (TENS)

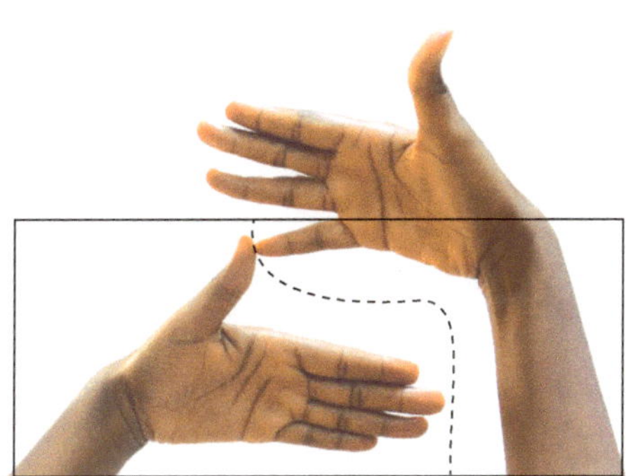

5 + 1 = 6

Place both numbers beside each other to obtain the answer.
10 × 6 = 60

Example 3.23

Solve 10 × 10.
Fingers touch each other.

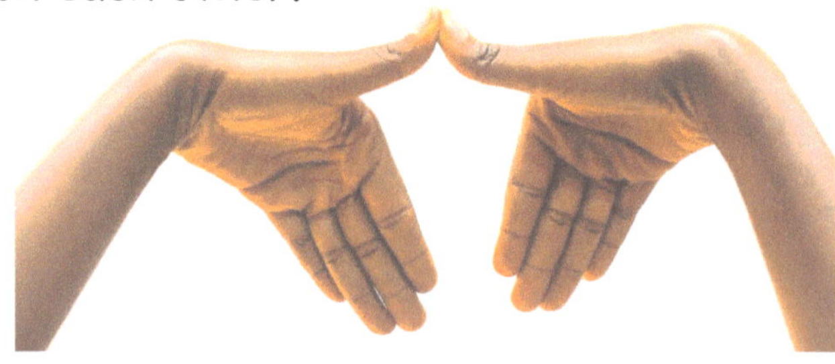

Count and multiply (UNITS)
0 × 0 = 0

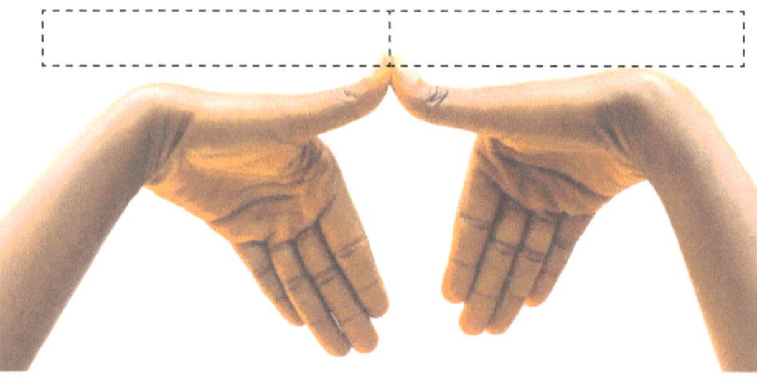

Count (TENS)

5 + 5 = 10

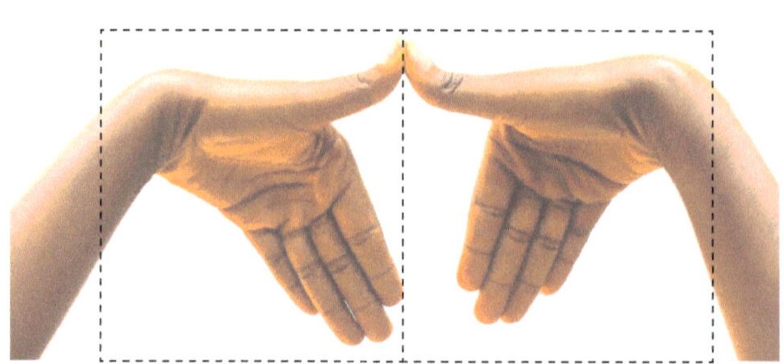

Place both numbers beside eachother to obtain the answer
10 × 10 = 100

Do It Yourself 8

1 10 × 7   **2** 10 × 8   **3** 10 × 9

It is essential to master elementary multiplication before attempting long multiplication.

This technique can be used to multiply numbers with several digits in no particular order.

Materials needed: pen or pencil and paper

Step 1: Split every number into digits and represent each digit with lines. For instance, if you have 52, draw 5 lines and 2 lines not 52 lines.

Step 2: Mark every point of intersection.

Step 3: Count diagonally.

Step 4: Write the number in a clockwise order. If the sum is more than one digit, add the Tens to the next number.

The line or graph multiplication technique can be applied to many cases including; **a)** One-digit numbers, **b)** Two-digit numbers, **c)** Three-digit numbers and **d)** Mixed numbers.

a) One-digit Numbers

Example 4.1

Solve 2 × 3.

Draw 2 lines diagonally.

Draw another 3 lines diagonally to intersect the 2 lines drawn previously.

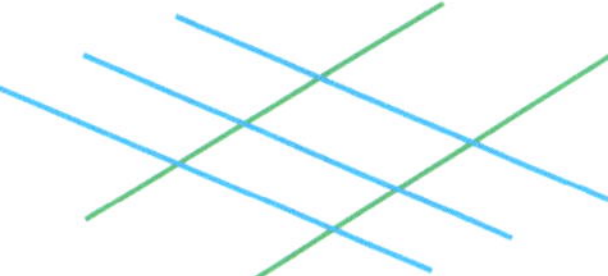

Mark points of intersection.

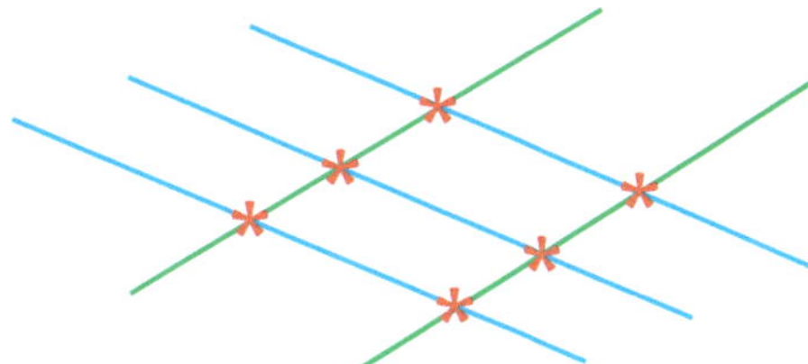

Count all the asterisks.
There are 6 points in all.
That is, 2 × 3 = 6

Solve 8 × 6.

Draw 8 lines diagonally, 6 lines to intersect it and asterisk all points of intersection.

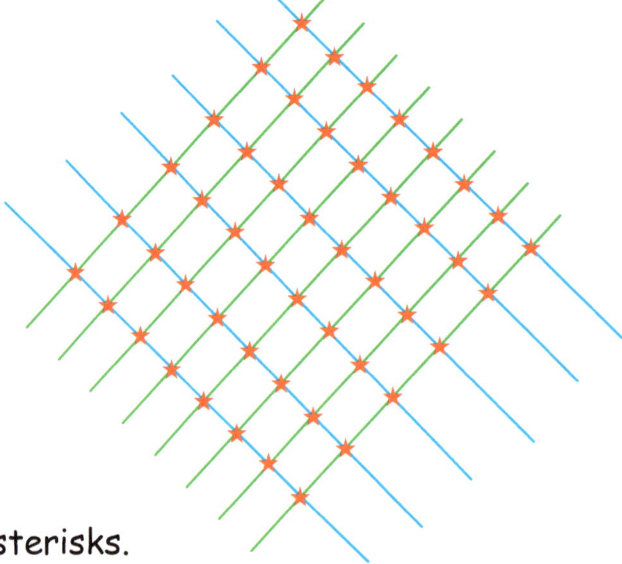

Count all the asterisks.
There are 48 in all.
Therefore, 8 × 6 = 48

Do It Yourself 9

With the aid of line multiplication, solve the following;

1 3 × 4   **2** 7 × 2   **3** 9 × 8

b) Two-digit Numbers

Two digits is slightly different from one digit in some ways;

i)   The number of lines will be the number of the digit not the whole number. For instance 21, will be 2 lines and 1 line drawn making three lines all together.

Please note that you are not to draw **21** lines.

ii) There will be more than one column which indicates that all the points of intersection will be counted column after column and the numbers written in a clockwise order.

Solve 21 × 13.

Draw 2 lines and 1 line diagonally with significant space between them.

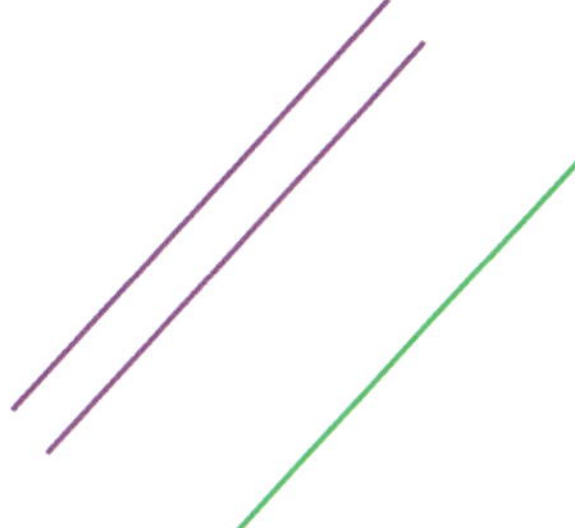

Draw 1 line and 3 lines diagonally to intersect.

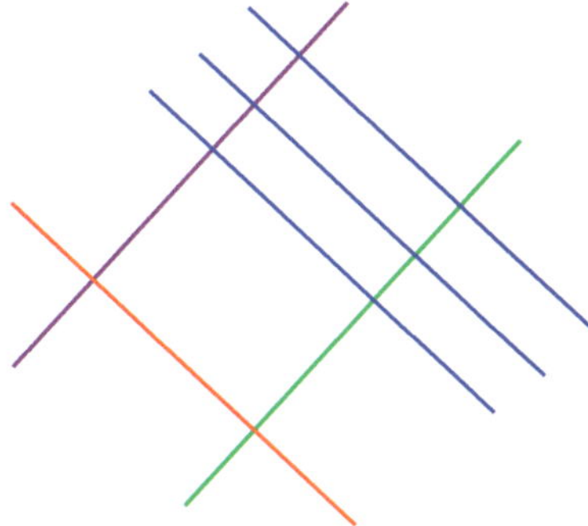

Mark all points of intersection.

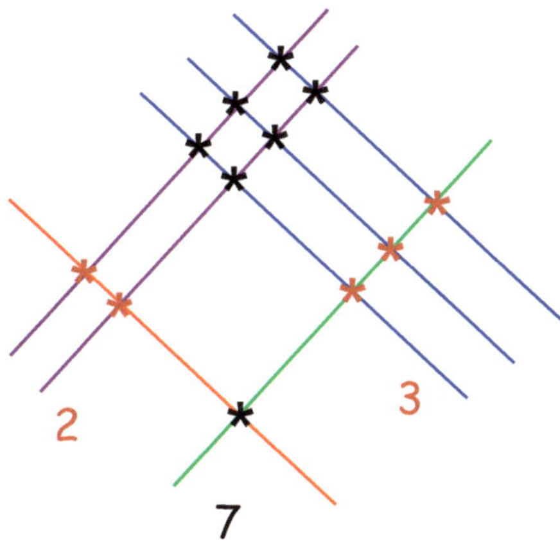

Count the points of intersection.

First diagonal: **2**

Second diagonal: **7**

Third diagonal: **3**

21 × 13 = 273

Solve 14 × 26.

Since question 1 was solved step-by-step, we will solve more steps at the same time here.

Draw 1 line and 4 lines diagonally with significant space between them.

Draw 2 line and 6 lines diagonally to intersect.

Mark all points of intersection

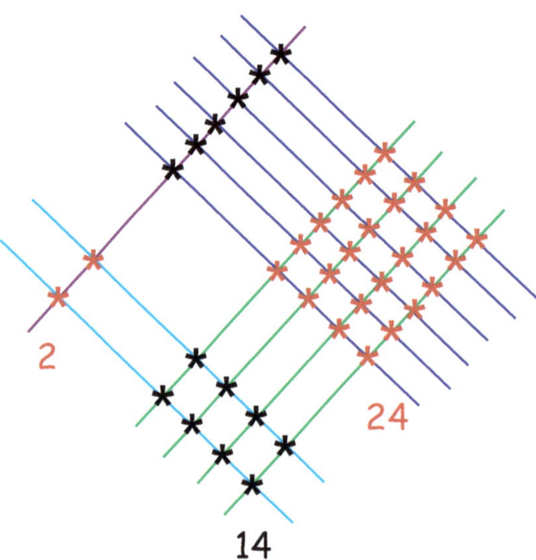

Count the points of intersection.

First diagonal: **2**

Second diagonal: **14**

Third diagonal: **24**

The points of intersection in the second and third diagonal are two digits, the answer is not 21424 rather add their TENS to the next number.

	H	T	U
	2		
+	1	4	
		2	4
	3	6	4

14 × 26 = 364

53 × 64

Draw **5** lines and **3** lines diagonally with significant space between them
Draw **6** lines and **4** lines diagonally to intersect.

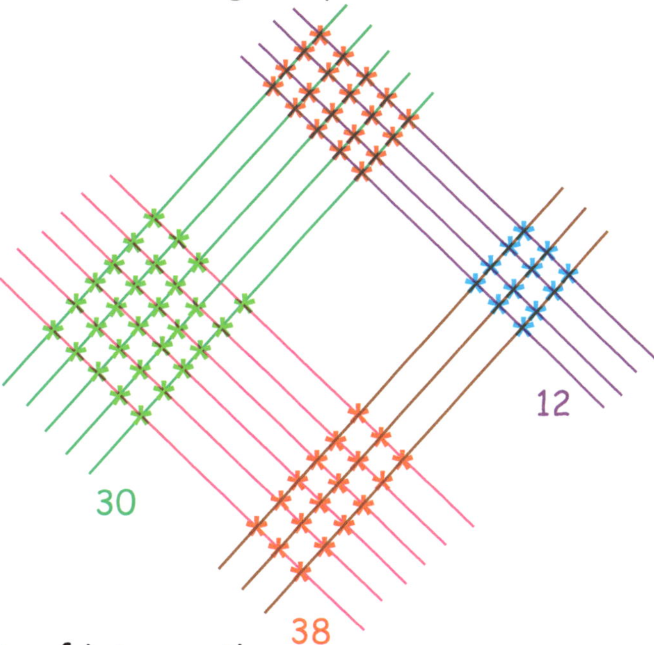

Mark all points of intersection.
Count the points of intersection.

First diagonal: **30**

Second diagonal: **38**

Third diagonal: **12**

	Th	H	T	U
	3	0		
+		3	8	
			1	2
	3	3	9	2

53 × 64 = 3392

Do It Yourself 10

1 Faderera bought 18 crates of soft drinks for her birthday party, how many bottles of soft drink does she have in total?
  (Note that each crate has 24 bottles)

c) Three-digit numbers

The principle applied to multiplying two digits can be applied here too

Example 4.6

Solve 121 × 113.

Draw line(s) to present each digit.

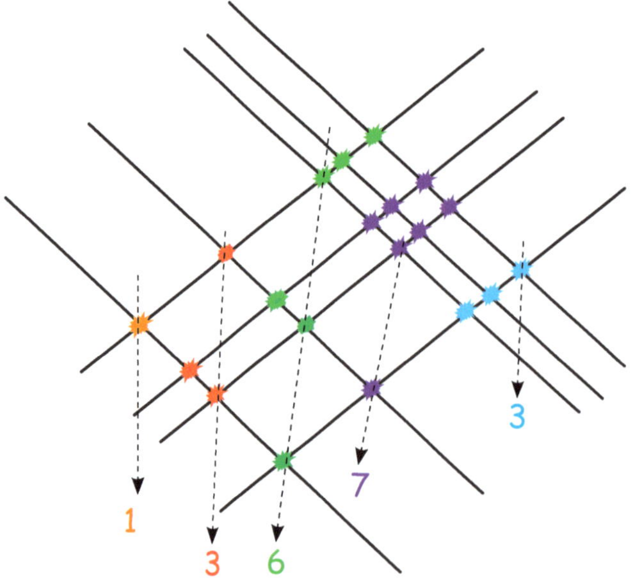

First diagonal: **1**
Second diagonal: **3**
Third diagonal: **6**
Fourth diagonal: **7**
Fifth diagonal: **3**

121 × 113 = 13673

Solve 133 × 421.

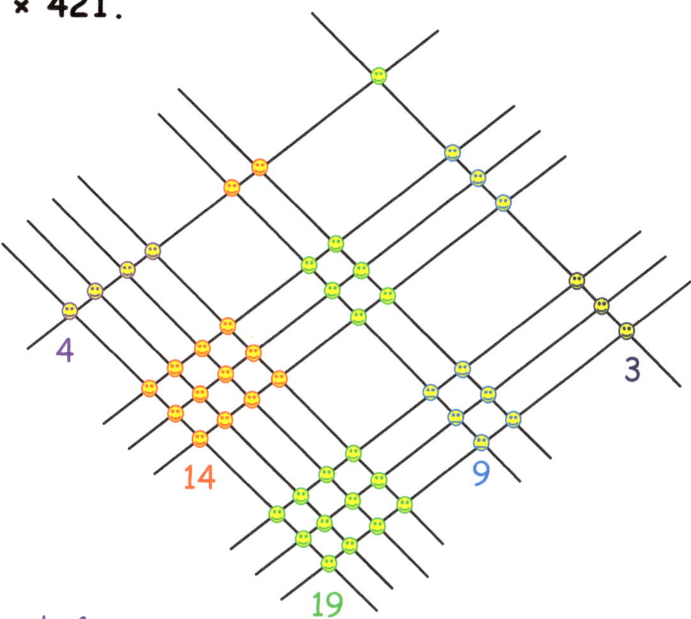

First diagonal: **4**
Second diagonal: **14**
Third diagonal: **19**
Fourth diagonal: **9**
Fifth diagonal: **3**

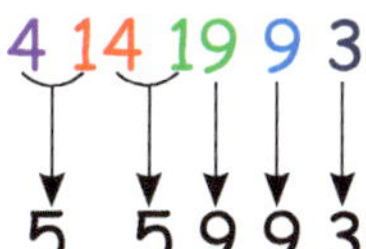

OR

TTh	Th	H	T	U
4				
1	4			
+	1	9		
			9	3
5	5	9	9	3

133 × 421 = 55993

323 × 241

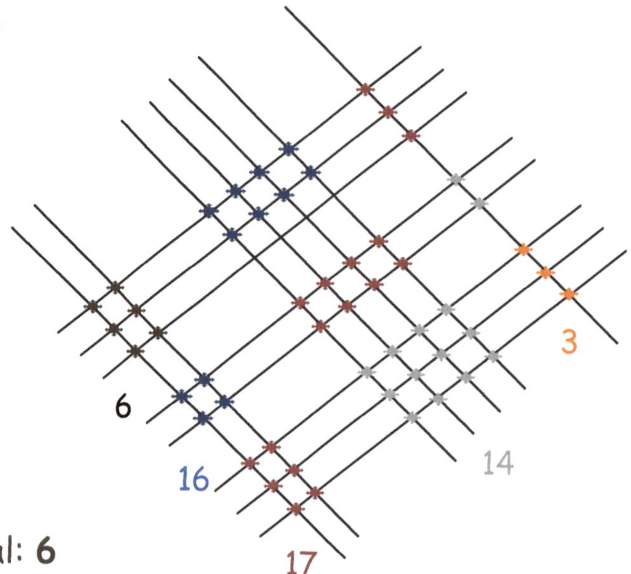

First diagonal: **6**
Second diagonal: **16**
Third diagonal: **17**
Fourth diagonal: **14**
Fifth diagonal: **3**

TTh	Th	H	T	U
6				
1	6			
	1	7		
+		1	4	
				3
7	7	8	4	3

323 × 241 = 77843

Do It Yourself 11

1 132 × 621
2 715 × 183
3 451 × 559

c) Mixed-digit numbers

Numbers with same digit(s) have been considered so far, we will now take on numbers with different digits.

Example 4.9

Solve 12 × 7.

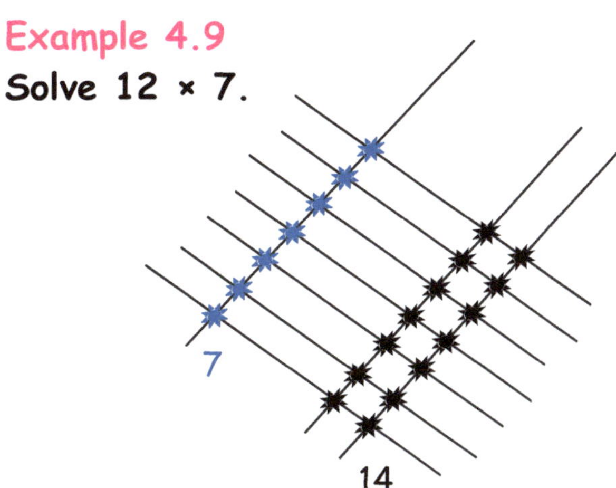

First diagonal: **7**

Second diagonal: **14**

	T	U
	7	
+ 1		4
	8	4

12 × 7 = 84

Example 4.10

Solve 236 × 4.

First diagonal: **8**

Second diagonal: **12**

Third diagonal: **24**

	H	T	U
	8		
+ 1		2	
		2	4
	9	4	4

$236 \times 4 = 944$

Example 4.11

67 × 3121

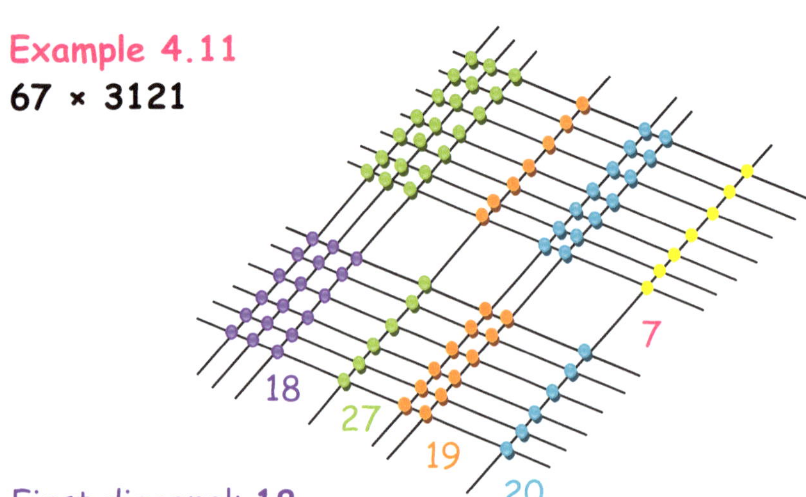

First diagonal: **18**
Second diagonal: **27**
Third diagonal: **19**
Fourth diagonal: **20**
Fifth diagonal: **7**

	HTh	TTh	Th	H	T	U
	1	8				
		2	7			
+			1	9		
				2	0	
						7
	2	0	9	1	0	7

$67 \times 3121 = 209107$

Do It Yourself 12

1 59 × 713   **2** 6341 × 24   **3** 421 × 2239

Numbers with zero

Since the digits are represented with lines, what happens when we try to multiply numbers with zero?

In this book, zero will be represented with dashes -----, instead of line(s) and when the dash(es) intersects a line or dash we do not count the point of intersection because they represent 0. However, when there are no marks in a diagonal, simply write 0 as the number in that diagonal. Another way to go about it is to mark all points of intersection and put a different sign on the intersections with dash(es). In this book we will be using this technique.

Example 4.12

40 × 202

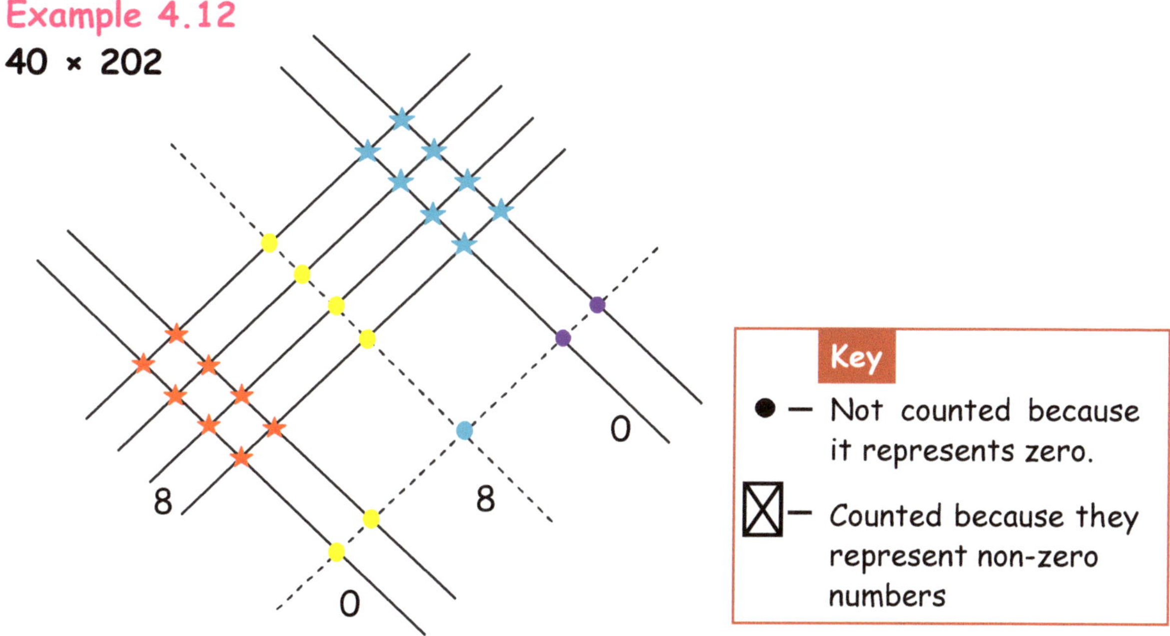

Points of line intersections were represented with ⊠ while points of dash intersections were represented with ● .

First diagonal: **8**

Second diagonal: 0+0=**0** (Remember that the circles represent 0.)

Third diagonal: 8+0=**8**

Fourth diagonal: **0**

40 × 202 = 8080

Example 4.13

Solve 101 × 304.

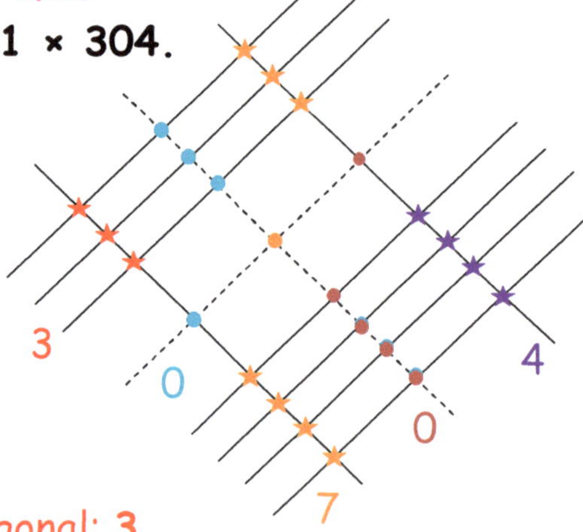

First diagonal: **3**

Second diagonal: 0 + 0 = **0** (Remember that the circles represent 0.)

Third diagonal: 3 + 0 + 4 = **7**

Fourth diagonal: 0 + 0 = **0**

Fifth diagonal: **4**

101 × 304 = 30704

Example 4.14

Solve 70 × 100.

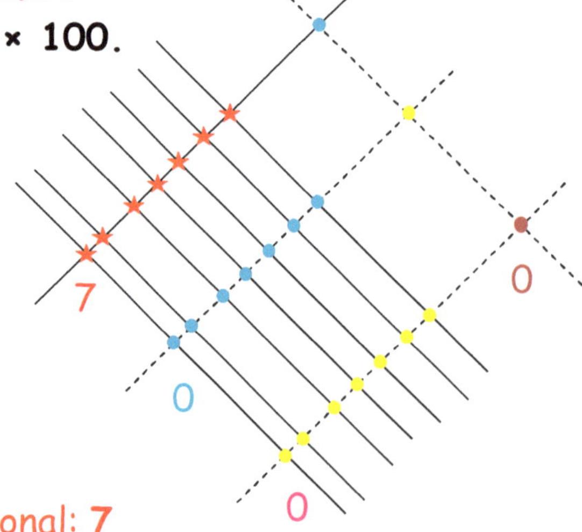

First diagonal: **7**

Second diagonal: 0 + 0 = **0** (Remember that the circles represent 0.)

Third diagonal: 0 + 0 = **0**

Fourth diagonal: **0**

70 × 100 = 7000

Do It Yourself 13

1 320 × 6   **2** 8440 × 13   **3** 2730 × 200

Chapter 5

Diamond Multiplication

Diamond multiplication technique can be used to multiply numbers with many digits, it is an alternative to the conventional long multiplication.

Materials Needed: pen and paper

Step 1: This technique involves drawing a diamond shape which will be partitioned into columns and rows; this is determined by the digits in the number under consideration.

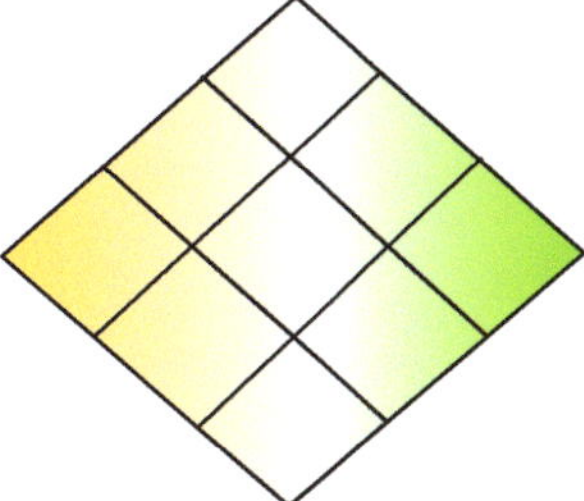

Step 2: Write the numbers on each column of the upper part of the diamond.

Step 3: Split each little diamond into two.

Step 4: Multiply the numbers that correspond to the two sides of each cell and write the answer in the cell while ensuring that each digit occupies the half of the cell earlier divided. If the result is one digit, place 0 infront of it. For instance, 4 × 2 = 8, however you write 08.

Step 5: Add all the numbers column after column. Should you obtain a two digit number in a column, remember to employ the addition technique earlier learnt.

Diamond multiplication technique can be applied to;

a) Same digits

b) Different digits

Same digits

Both numbers multiplied have equal digits.

Example 5.1

12 × 35

Since both numbers are two digits each, draw a 2 × 2 (called 2 by 2) diamond.

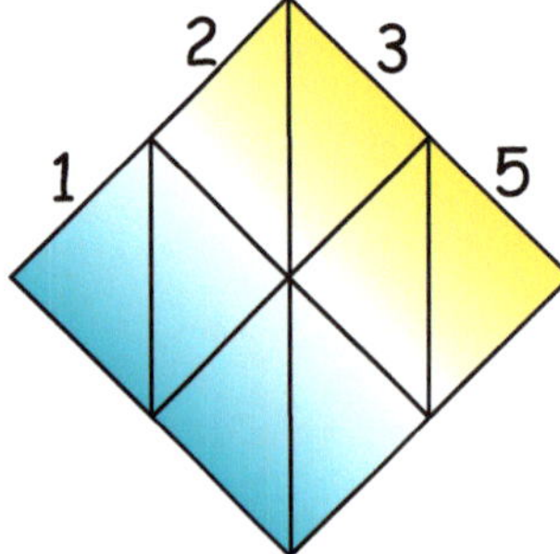

Split each little diamond into two equal **halves** and write the numbers to be multiplied on each column.

Fill the cells by multiplying;

1 × 3 = 03

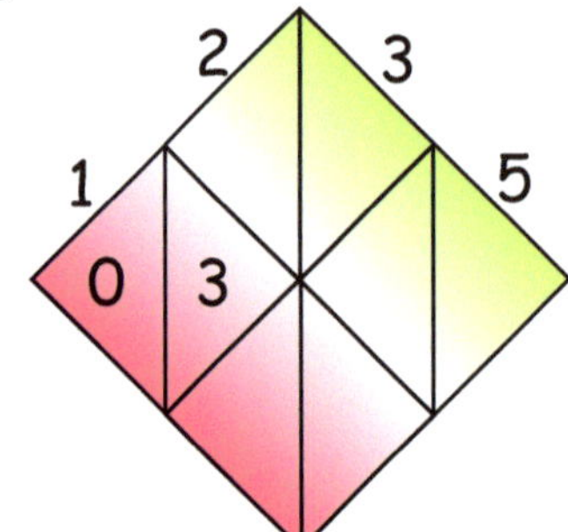

2 × 3 = 06

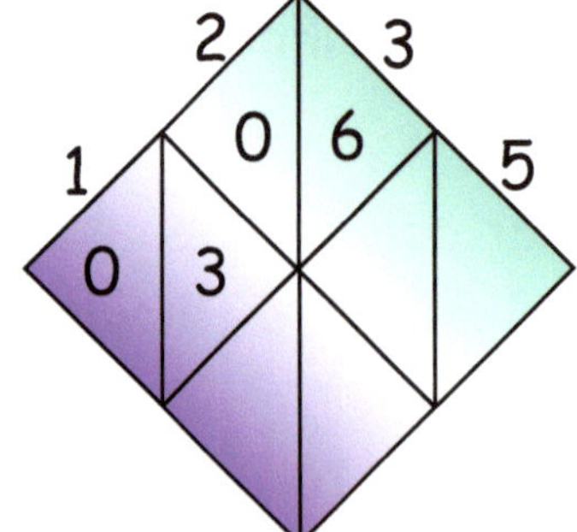

1 × 5 = 05

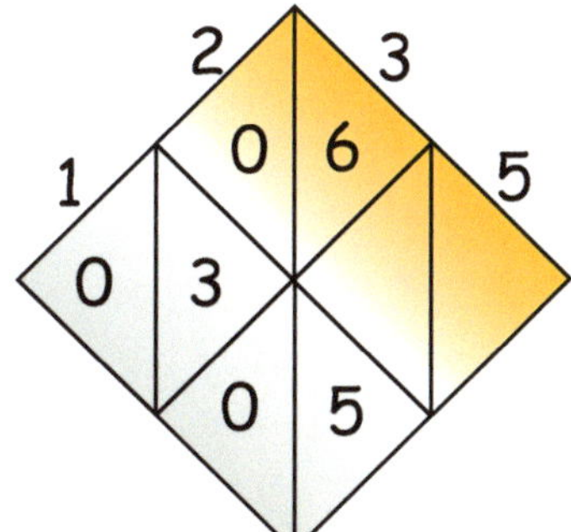

2 × 5 = 10

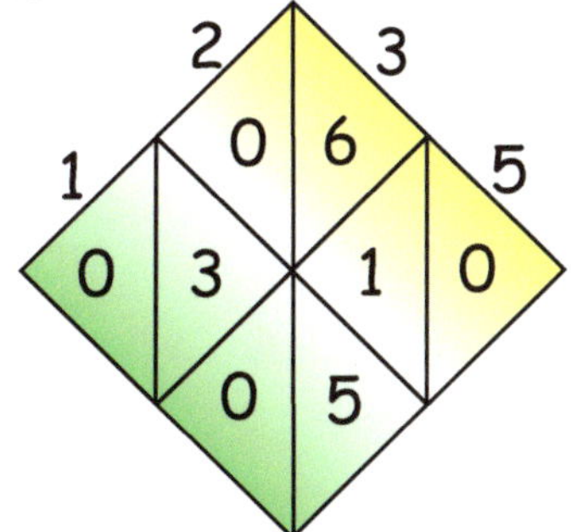

Add each column.

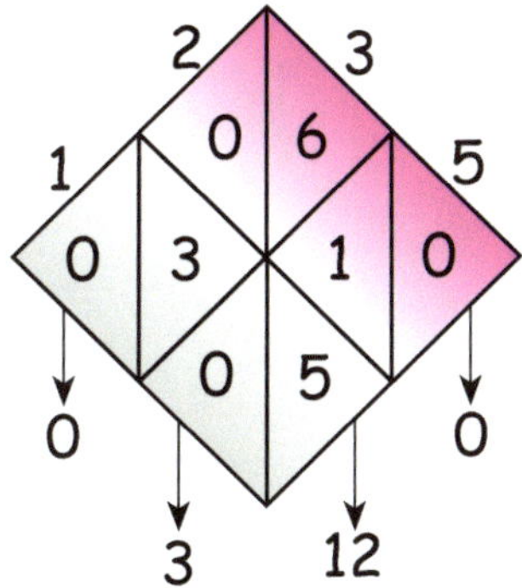

H	T	U
3		
+ 1	2	
		0
4	2	0

Therefore, 12 × 35 = 420

345 × 261

Since both numbers are three digits each, draw a 3×3 (called 3 by 3) diamond.

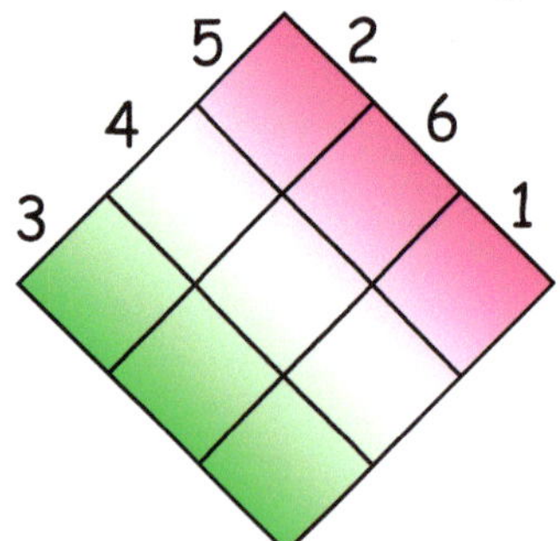

Split each little diamond into two equal **halves** and write the numbers to be multiplied on each column.

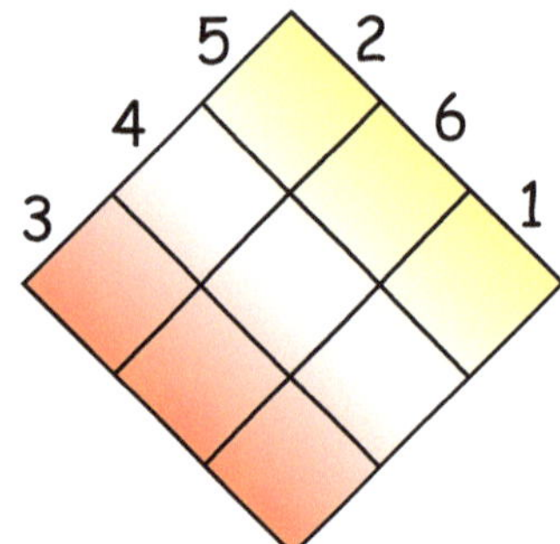

Fill the cells by multiplying;

3 × 2 = 06

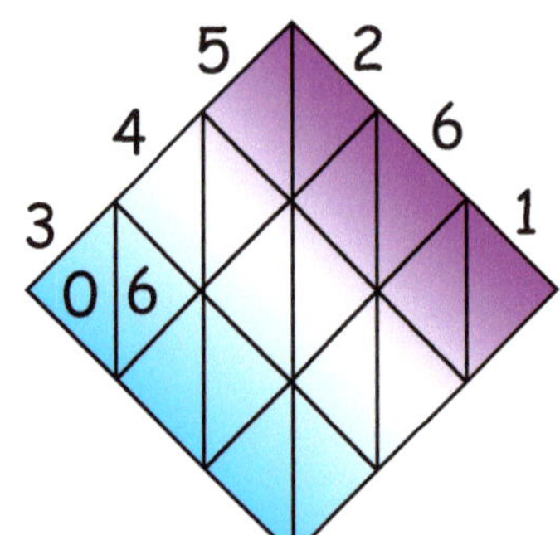

4 × 2 = 08

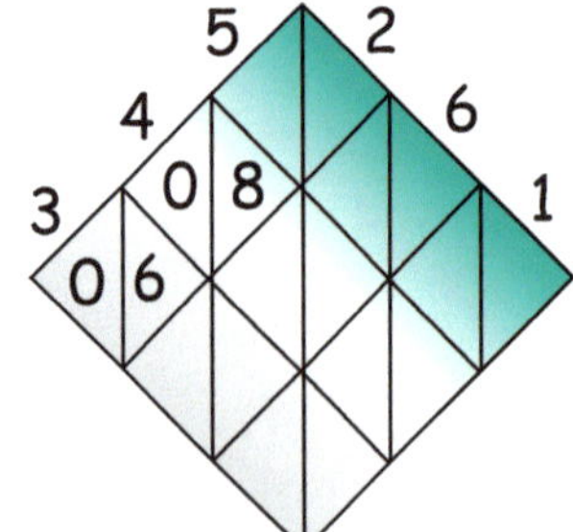

5 × 2 = 10

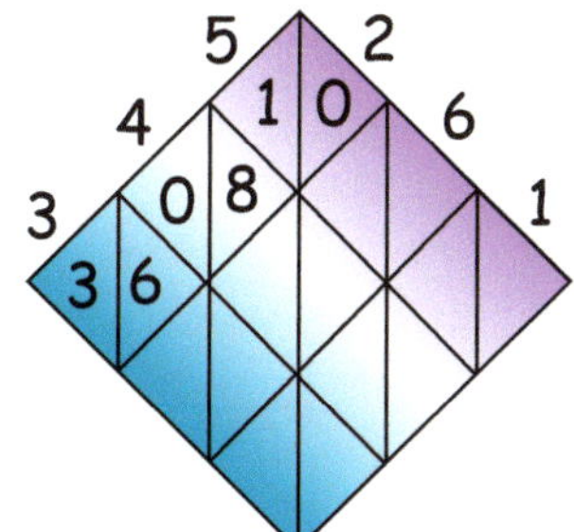

3 × 6 = 18

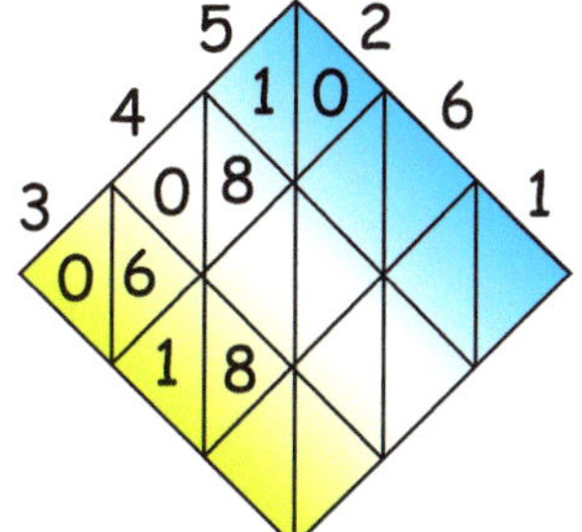

4 × 6 = 24

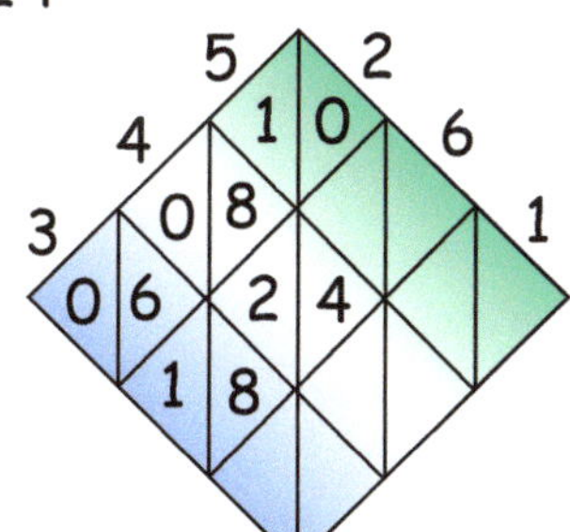

5 × 6 = 30

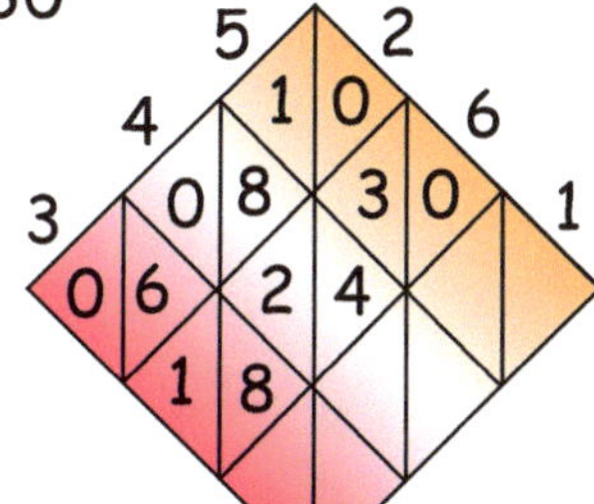

3 × 1 = 03

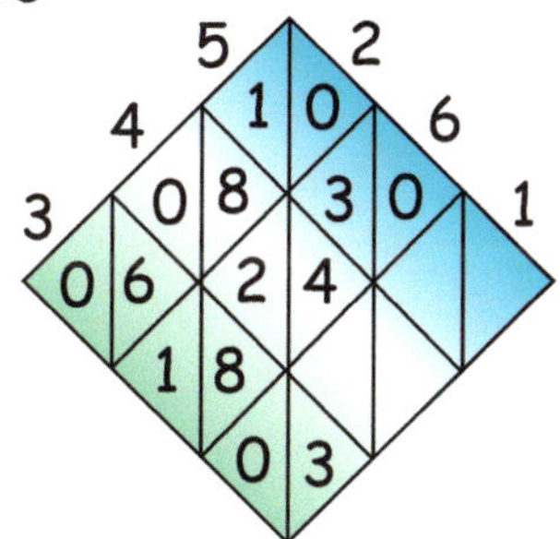

4 × 1 = 04

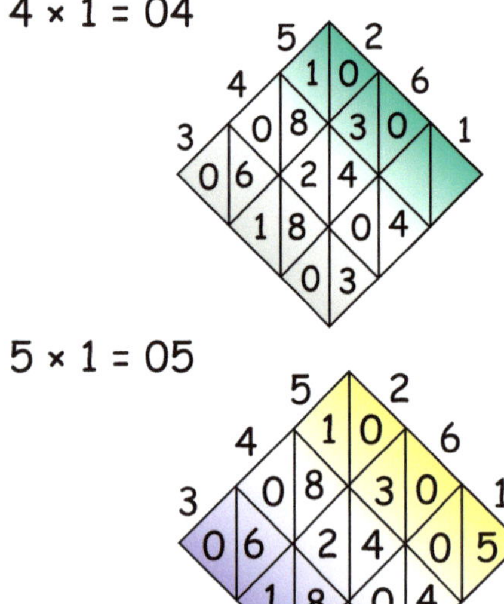

5 × 1 = 05

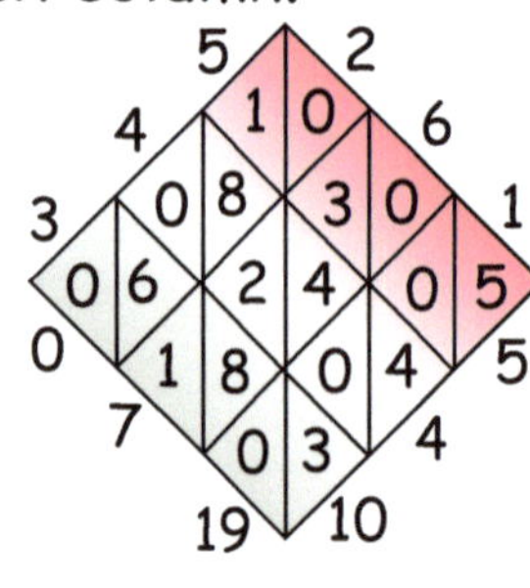

Add each column.

HTh	TTh	Th	H	T	U
0					
	7				
	1	9			
+		1	0		
				4	
					5
0	9	0	0	4	5

Therefore, 345 × 261 = 90045

Using diamond multiplication, solve the following.

1 26 × 41   **2** 345 × 822   **3** 4351 × 2823

Different digits

Both numbers multiplied do not have same digits.

Example 5.3

147 × 6

Since one number has three digits and the other number has one digit each, draw a 3 × 1 diamond.

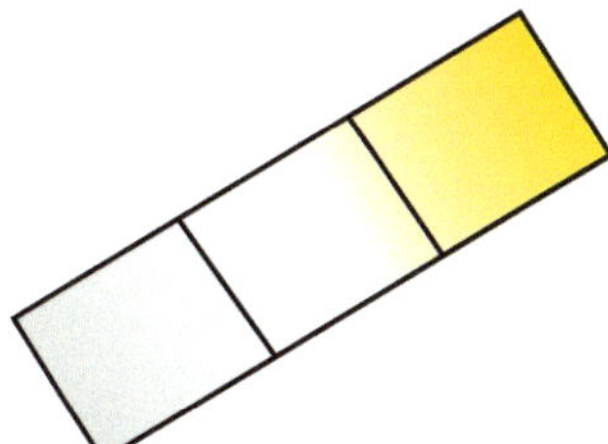

Split each little diamond into two equal halves and write the numbers to be multiplied on each column.

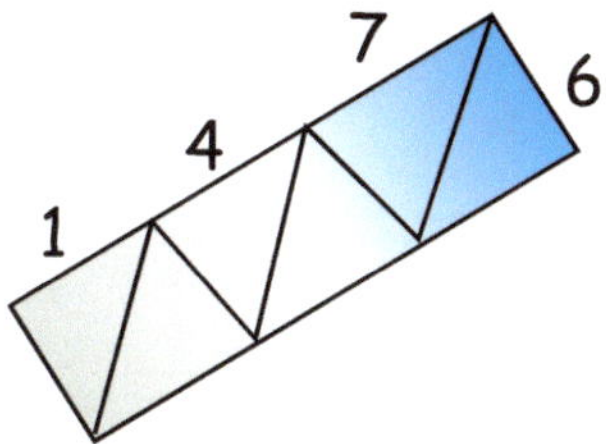

Fill the cells by multiplying;

1 × 6 = 06

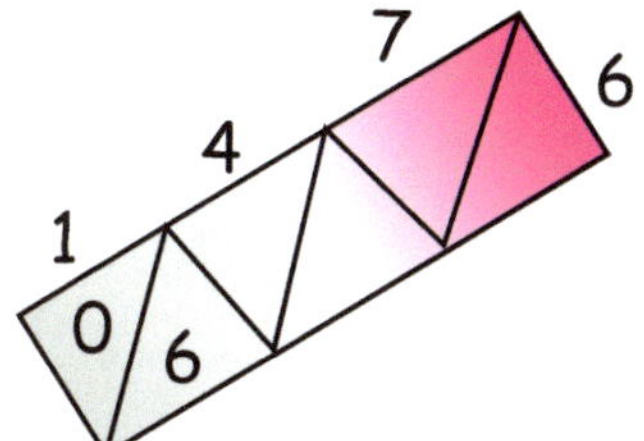

4 × 6 = 24

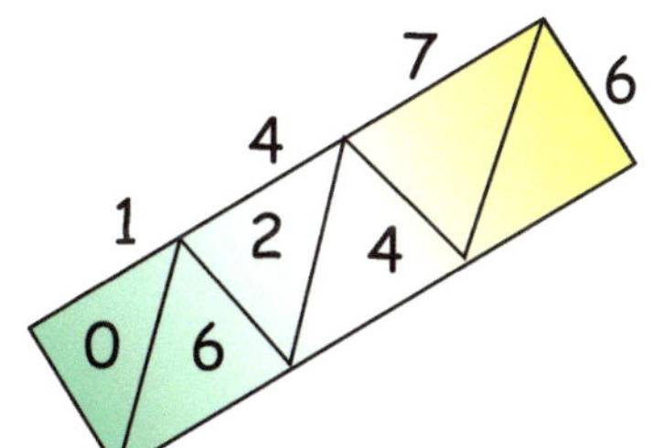

7 × 6 = 42

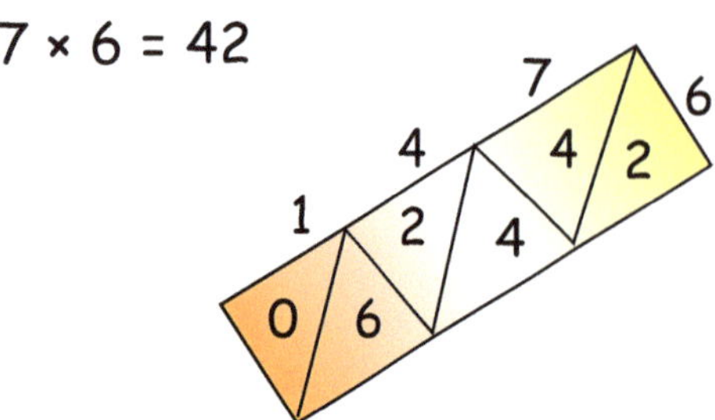

Add each column.

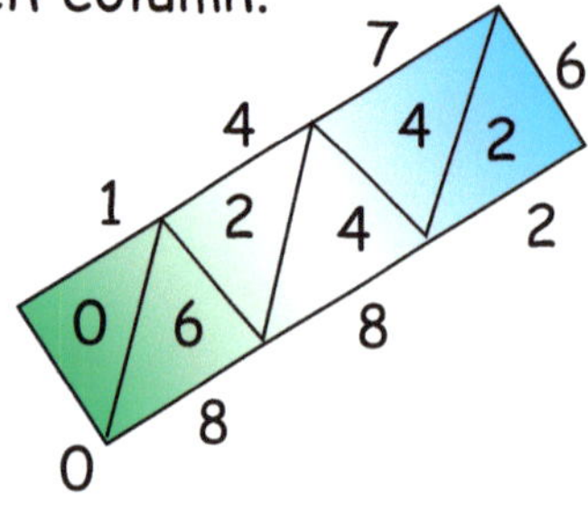

147 × 6 = 882

Example 5.4

22 × 523

Since one number has two digits and the other number has three digits each, draw a 2 × 3 diamond.

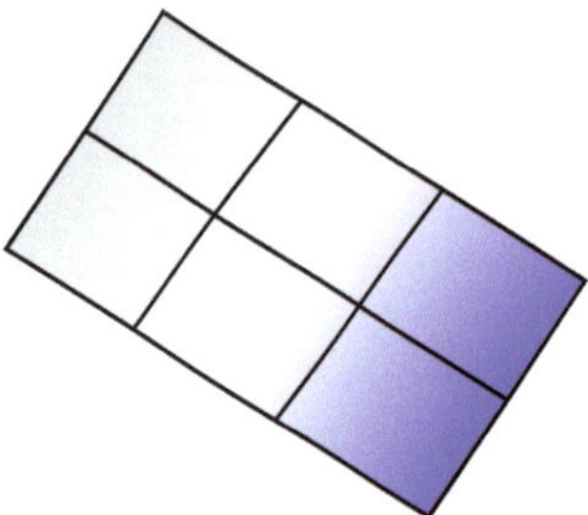

Split each little diamond into two equal **halves** and write the numbers to be multiplied on each column.

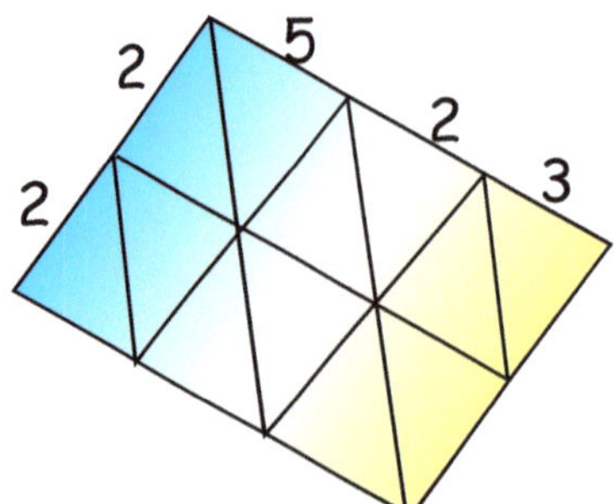

Fill the cells by multiplying;

2 × 5 = 10

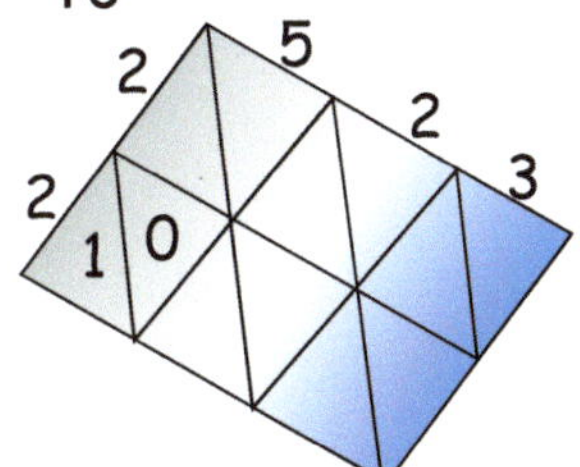

2 × 5 = 10

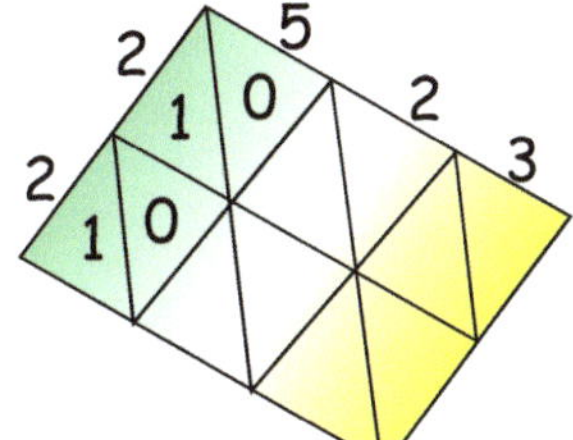

2 × 2 = 04

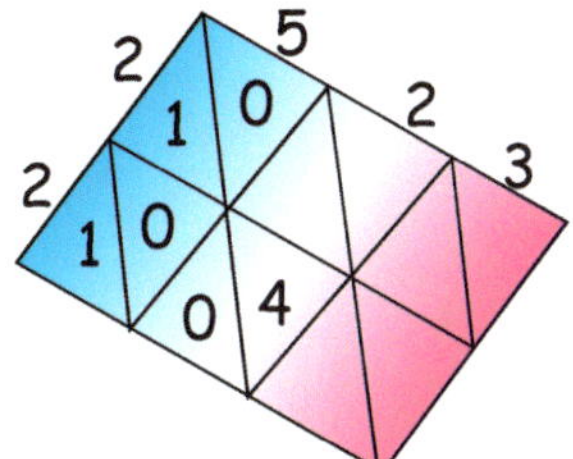

2 × 2 = 04

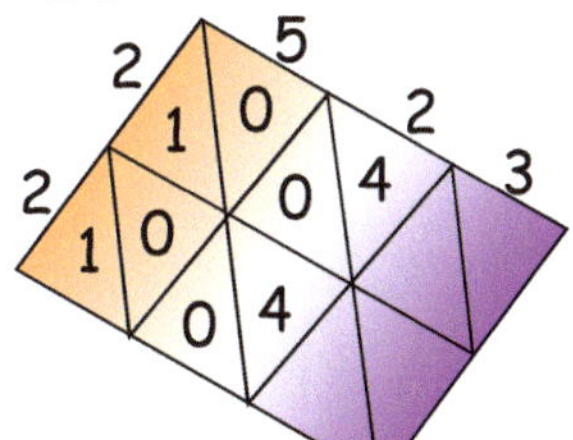

2 × 3 = 06

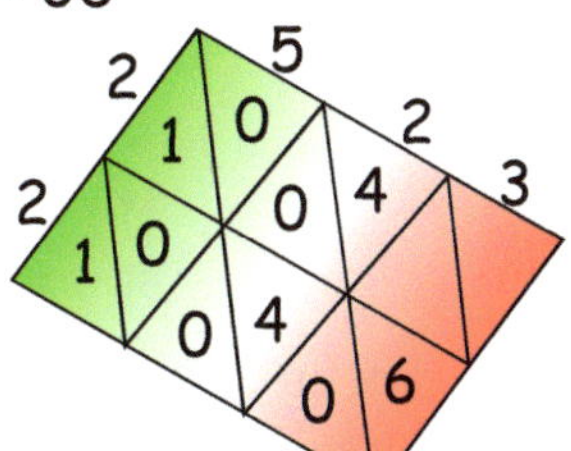

2 × 3 = 06

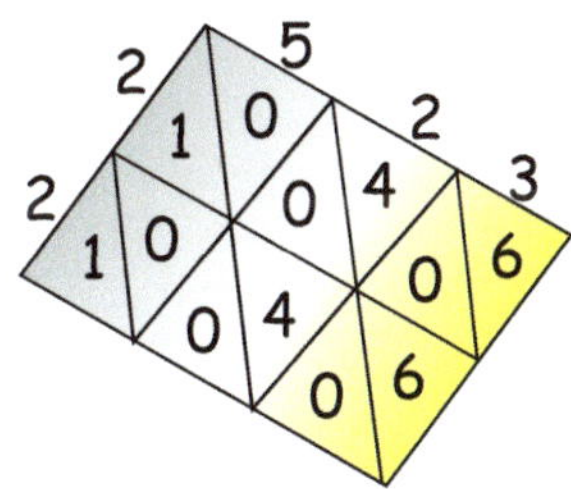

Add each column.

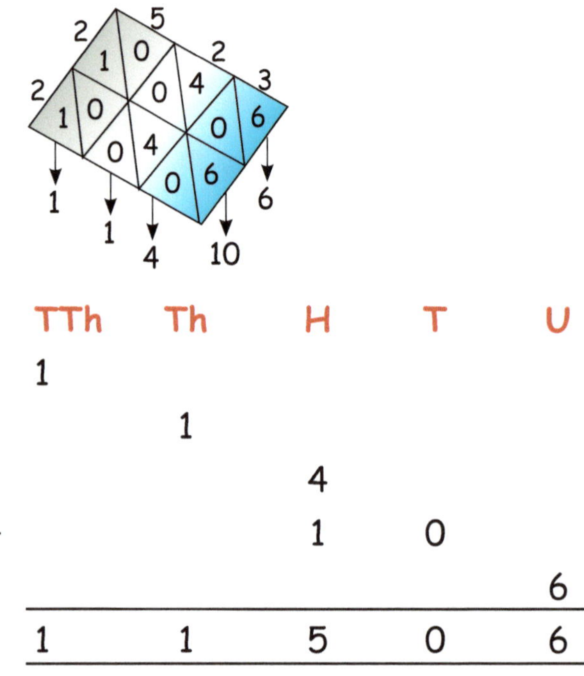

TTh	Th	H	T	U
1				
	1			
		4		
+		1	0	
				6
1	1	5	0	6

Therefore, 22 × 523 = 11506

Example 5.5

93 × 1324

Since one number has two digits and the other number has four digits each, draw a 2 × 4 diamond.

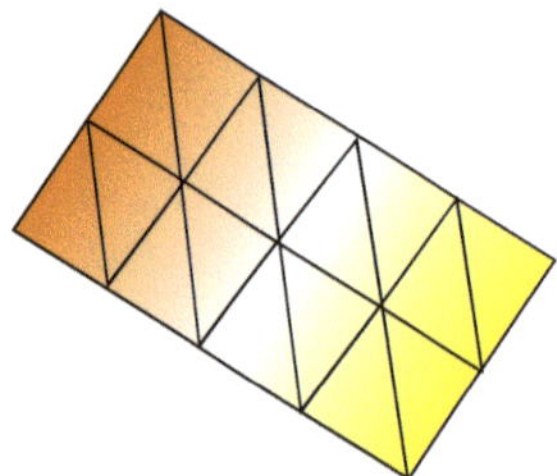

Split each little diamond into two equal **halves** and write the numbers to be multiplied on each column.

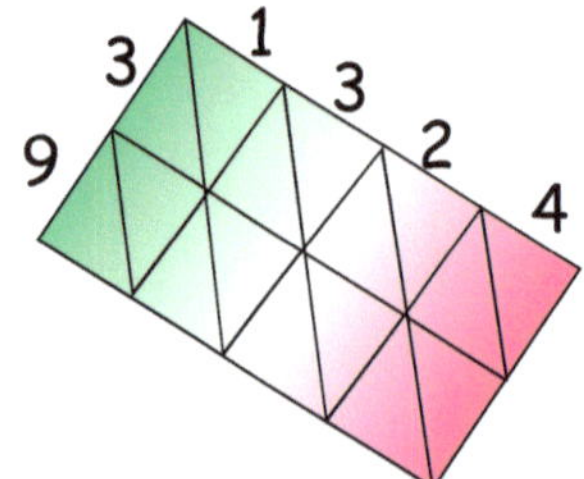

Fill the cells by multiplying;

9 × 1 = 09

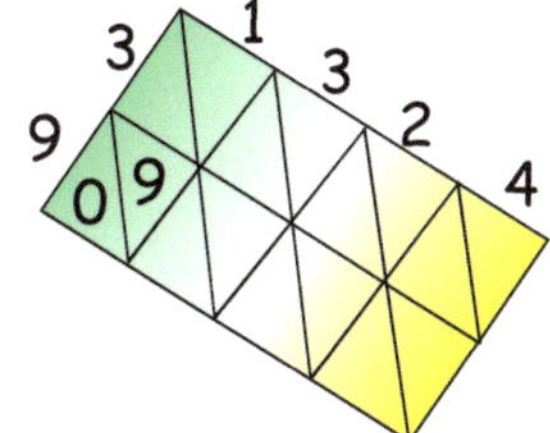

3 × 1 = 03

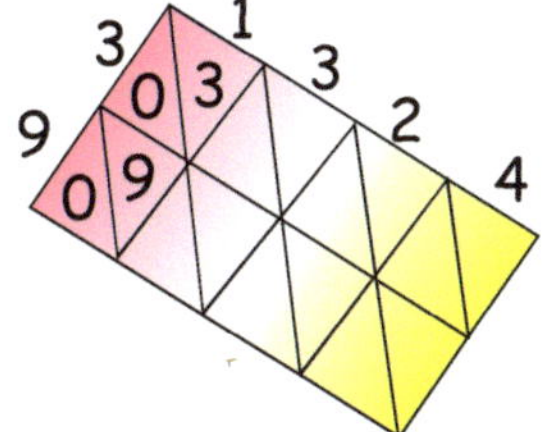

9 × 3 = 27

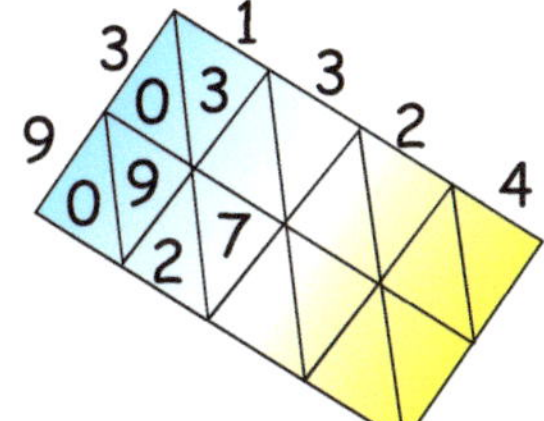

3 × 3 = 09

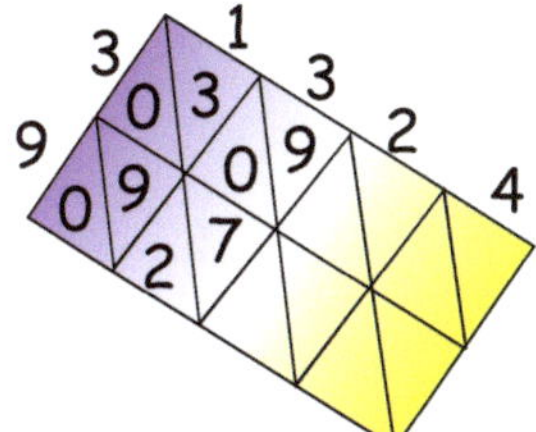

9 × 2 = 18

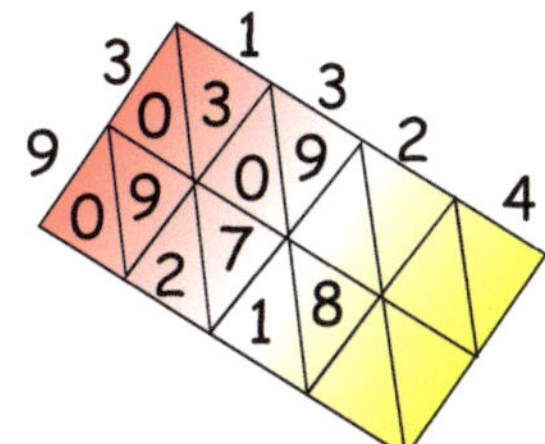

3 × 2 = 06

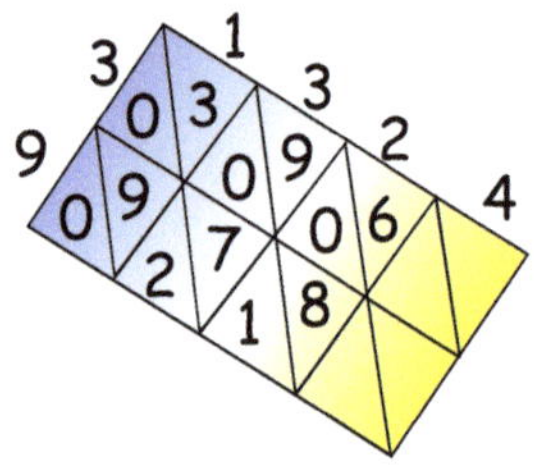

9 × 4 = 36

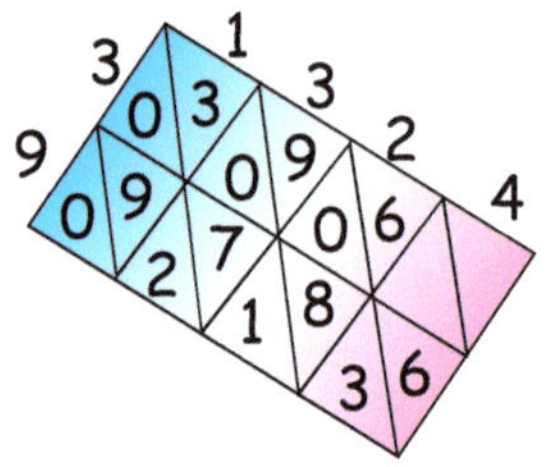

3 × 4 = 12

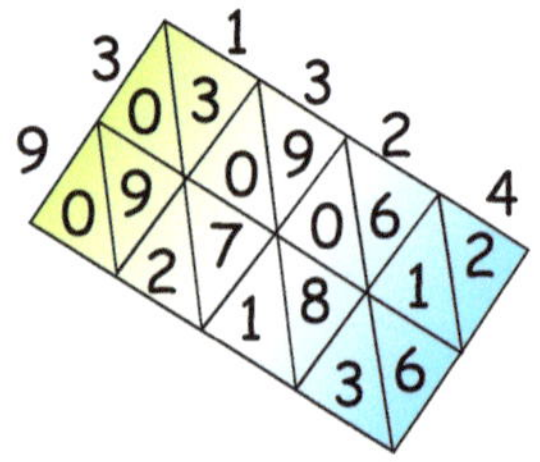

Add each column.

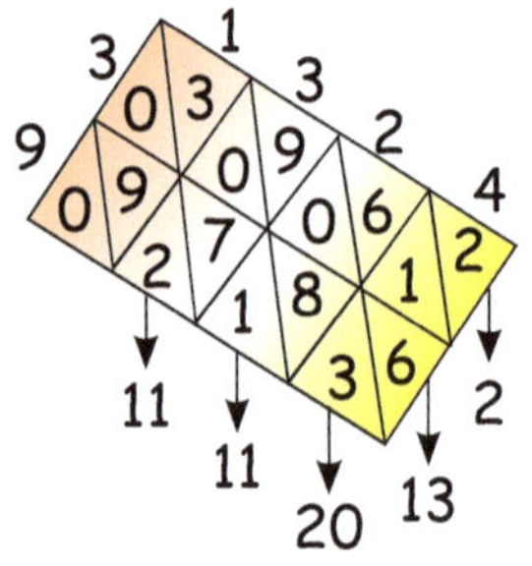

	HTh	TTh	Th	H	T	U
	1	1				
		1	1			
			2	0		
+				1	3	
						2
	1	2	3	1	3	2

93 × 1324 = 123132

Do It Yourself 15

Solve the following with Diamond multiplication method.

1 9 × 16 **2** 27 × 823 **3** 63846 × 572

Chapter 6

B□x Multiplication

Box multiplication is an alternative substitute for long multiplication. Remember the aim of this book is to ensure you learn how to multiply with ease. When you see the technique that suits you most, simply stick to it.

Materials Needed: Pen and paper

Step 1: Split each number into their various place value systems.

Step 2: Draw a table whose dimension will be determined by the numbers to be multiplied and write the numbers on the topmost row and first column.

To determine the dimension of the matrix, add one to the total digit(s) of each number.

For instance, if you are to multiply 45 by 678; 45 has two digits 4 and 5, add one to the two digits so you have three also 678 has three digits 6, 7 and 8, add one to the three digits to have four. The dimension, therefore, will be 3 by 4.

It is pertinent to note that the order matter less due to the commutative nature of multiplication. As regards to the example just considered, the dimension can be 3 × 4 or 4 × 3, either way will give same result.

Step 3: Multiply the number at the top of each cell by the number on the first column on the left

Step 4: Sum every number.

Examples 6.1

32 × 79

Split:

32 ▯ 30 + 2

79 ▯ 70 + 9

Since both numbers are two digits each, draw a 3 × 3 matrix.

✕	30	2
70		
9		

Multiply the number on the topmost cell by the number on the left, now let us do this one after the other;

70 × 30 = 2100

×	30	2
70	2 100	
9		

70 × 2 = 140

×	30	2
70	2 100	140
9		

9 × 30 = 270

×	30	2
70	2 100	140
9	270	

9 × 2 = 18

×	30	2
70	2 100	140
9	270	18

Sum every number

Th	H	T	U
2	1	0	0
	1	4	0
+	2	7	0
		1	8
2	5	2	8

32 × 79 = 2 528

607 × 342

Split:

607 ⧇ 600 + 0 + 7

342 ⧇ 300 + 40 + 2

Since both numbers are three digits each, draw a 4 × 4 matrix. Remember you add one to the total digit(s) of each number to determine the dimension.

×	600	0	7
300			
40			
2			

Fill the boxes/cells by multiplying the numbers row by column

300 × 600 = 180 000

×	600	0	7
300	180 000		
40			
2			

300 × 0 = 0

×	600	0	7
300	180 000	0	
40			
2			

300 × 7 = 2 100

×	600	0	7
300	180 000	0	2 100
40			
2			

40 × 600 = 24 000

×	600	0	7
300	180 000	0	2 100
40	24 000		
2			

40 × 0 = 0

×	600	0	7
300	180 000	0	2 100
40	24 000	0	
2			

40 × 7 = 280

×	600	0	7
300	180 000	0	2 100
40	24 000	0	280
2			

2 × 600 = 1 200

×	600	0	7
300	180 000	0	2 100
40	24 000	0	280
2	1 200		

2 × 0 = 0

×	600	0	7
300	180 000	0	2 100
40	24 000	0	280
2	1 200	0	

2 × 7 = 14

×	600	0	7
300	180 000	0	2 100
40	24 000	0	280
2	1 200	0	14

Sum every number.

HTh	TTh	Th	H	T	U
1	8	0	0	0	0
					0
		2	1	0	0
	2	4	0	0	0
					0
			2	8	0
+			1	2	0	0
					0
				1	4
2	0	7	5	9	4

607 × 342 = 207 594

Examples 6.3

7 671 × 452

Split:

7 671 ☒ 7 000 + 600 + 70 + 1

The first number 7 671 has four digits (7,6,7,1) add one to four to have **five**; the second number 452 has three digits (4,5,2) add one to three to have **four**. The dimension then is 5 × 4 (5 by 4).

Draw a 5 × 4 matrix.

×	7 000	600	70	1
400				
50				
2				

Fill the boxes one after the other.
400 × 7 000 = 2 800 000

×	7 000	600	70	1
400	2 800 000			
50				
2				

400 × 600 = 240 000

×	7 000	600	70	1
400	2 800 000	240 000		
50				
2				

400 × 70 = 28 000

×	7 000	600	70	1
400	2 800 000	240 000	28 000	
50				
2				

$400 \times 1 = 400$

×	7 000	600	70	1
400	2 800 000	240 000	28 000	400
50				
2				

$50 \times 7\ 000 = 350\ 000$

×	7 000	600	70	1
400	2 800 000	240 000	28 000	400
50	350 000			
2				

$50 \times 600 = 30\ 000$

×	7 000	600	70	1
400	2 800 000	240 000	28 000	400
50	350 000	30 000		
2				

$50 \times 70 = 3\ 500$

×	7 000	600	70	1
400	2 800 000	240 000	28 000	400
50	350 000	30 000	3 500	
2				

$50 \times 1 = 50$

×	7 000	600	70	1
400	2 800 000	240 000	28 000	400
50	350 000	30 000	3 500	50
2				

2 × 7000 = 14 000

×	7 000	600	70	1
400	2 800 000	240 000	28 000	400
50	350 000	30 000	3 500	50
2	14 000			

2 × 600 1 200

×	7 000	600	70	1
400	2 800 000	240 000	28 000	400
50	350 000	30 000	3 500	50
2	14 000	1 200		

2 × 70 = 140

×	7 000	600	70	1
400	2 800 000	240 000	28 000	400
50	350 000	30 000	3 500	50
2	14 000	1 200	140	

2 × 1 = 2

×	7 000	600	70	1
400	2 800 000	240 000	28 000	400
50	350 000	30 000	3 500	50
2	14 000	1 200	140	2

Sum

M	HTh	TTh	Th	H	T	U
2	8	0	0	0	0	0
	2	4	0	0	0	0
		2	8	0	0	0
				4	0	0
	3	5	0	0	0	0
		3	0	0	0	0
			3	5	0	0
					5	0
		1	4	0	0	0
			1	2	0	0
+				1	4	0
						2
3	4	6	7	2	9	2

7671 × 452 = 3467292

Do It Yourself 16

1 Toluwani orders 176 crates of soft drinks for her birthday party, how many bottles will she have in total considering that each crate contains 24 bottles?

2 Chioma got 928 packs of candies for her friends, each pack contains 115 pieces. How many candies does she have altogether?

Assuredly, various means of multiplying considered earlier has been mind blowing as they are things we probably never knew or thought about till now.

Imagine yourself shopping or in any public place and you need to multiply quickly, you may not want to count your fingers (body part) publicly or request for paper or pen to draw lines, to this end, mental multiplication is almost inevitable in our everyday dealings.

There are several principles involved in mental multiplication. This should not discourage you but rather challenge you to practise till you have mastered them all. Remember, it is constant practise that births perfection.

The advantage of practising is that it enhances the capacity of your brain which will help in decision making; career-wise and even academically.

Mental multiplication consumes lesser time. The only material needed is the brain.

The techniques to multiply mentally include:

1 Base multiplication   2 Rainbow multiplication   3 Double-Half

Base Multiplication

Base multiplication involves making a number your reference point. It is important to know that not all numbers can be used as reference points (base).

Base 10

The reference point is 10. This can be used to multiply numbers below ten and slightly above ten. However, this method cannot be used to multiply any number by 10 i.e. multiples of ten are excluded such as 4×10, 7×10, 15×10. This is no difficult task since all we need to do to multiply a number by 10 is place '0' behind it. For instance, $4 \times 10 = 40$, $7 \times 10 = 70$, $15 \times 10 = 150$.

Numbers below 10

The numbers below 10 include 1, 2, 3,..., 9. Since the smaller numbers are easily known by heart, this technique is recommended for higher numbers between 6 and 9.

Steps

A × B

1   Subtract both numbers from 10:10 – A = C

$$10 – B = D$$

2   Subtract the answer from the other number:

10 – A = C

10 – B = D

That is,

B – C = E

A – D = E

Answer obtained for both will be the same.

E will represent **TENS**.

3   Multiply the two values you got initially C × D = F.

F will represent **UNIT**.

4   Therefore, A × B = EF

NOTE: If the result of C × D is a two digit number. i.e. C × D = GF, add G to F

A × B = EGF

        = (E + G) F

        = HF

Example 7.1

Solve 6 × 5.

Step 1: Subtract both numbers from 10 : 10 – 6 = 4

$$10 – 5 = 5$$

Step 2: Subtract the answer from the other number:

10 – 6 = 4

10 – 5 = 5

That is,

6 – 5 = 1

5 – 4 = 1

'1' will represent **TENS**.

Step 3: Multiply the two values you got in **step 1**

4 × 5 = 20.

'20' will represent **UNIT**.

Step 4:

```
     T        U
     1
 +   2        0
 ─────────────────
     3        0
 ─────────────────
   6 × 5 = 30
```

Example 7.2

Solve 4 × 9.

Step 1: Subtract both numbers from 10 : 10 – 4 = 6

10 – 9 = 1

Step 2: Subtract the answer from the other number:

10 – 4 = 6

10 – 9 = 1

That is,

9 – 6 = 3

4 – 1 = 3

'3' will represent **TENS**.

Step 3: Multiply the two values you got in **Step 1**.

6 × 1 = 6.

'6' will represent **UNIT**.

Step 4:

```
     T       U
     3
 +           6
 ___________
     3       6
 ___________
```

$$4 \times 9 = 36$$

Example 7.3

Solve 7 × 8.

Step 1: $10 - 7 = 3$
$10 - 8 = 2$

Step 2: $10 - 7 = 3$

$10 - 8 = 2$

That is,
$8 - 3 = 5$
$7 - 2 = 5$
'5' **(TENS)**

Step 3: $3 \times 2 = 6$
'6' **(UNIT)**

Step 4:

```
     T       U
     5
 +           6
 ___________
     5       6
 ___________
```

$$7 \times 8 = 56$$

Do It Yourself 17

Multiply the following using base multiplication method:

1 5 × 8　　　　**2** 6 × 9　　　　**3** 9 × 7

Numbers above 10

Technically, whole numbers above ten are from 11 to infinity, however, this method will be restricted to 11 × 11...19 × 19.

Steps

AB × CD

1   Subtract 10 from one of the numbers AB – 10 = E

or
CD – 10 = F

2   Add the result to the other number

E – CD = G

 or

F + AB = G

G will represent **TENS**.

3   Multiply the last digit of the two numbers been multiplied B × D = H.
F will represent **UNIT**.

4   Therefore, AB × CD = GH

Example 7.4

1   **13 × 12 = 30**

Step 1: Subtract 10 from one of the numbers

13 – 10 = 3 or 12 – 10 = 2

Step 2: Add the result to the other number

3 + 12 = 15 or 2 + 13 = 15

15 will represent **TENS**.

Step 3: Multiply the last digit of the two numbers been multiplied

3 × 2 = 6

6 will represent **UNIT**.

Therefore, 13 × 12 = 156

2 16 × 15

Step 1: 16 – 10 = 6   or   15 – 10 = 5
Step 2: 6 + 15 = 21 **(TENS)**   or   5 + 16 = 21 **(TENS)**
Step 3: 6 × 5 = 30 **(UNIT)**

H	T	U
2	1	
	3	0
2	4	0

Therefore, 16 × 15 = 240

Do It Yourself 18

With the aid of a Base Multiplication method, multiply the following;

1   11 × 14   2   13 × 15   3   17 × 18

Numbers above 19

What happens to numbers above 19?
The procedure is similar to the above however some additional steps will be taken.

Steps

AB × CD

1   Subtract one of the numbers from their respective TENS. For instance, the TENS of 26 is 20, 65 is 60, 82 is 80.
     AB – TENS = E, CD – TENS = F
2   Add the result to the other number i.e.
     E + CD = G or F + AB = G
3   Multiply G with the TENS of the multiplied number i.e.
     G × TENS = HI (Hundred Tens)
4   Multiply the last digits of both numbers been multiplied
     B × D = J (Units)
5   AB × CD = HIJ

Example 7.5

1   23 × 28 = 30

Step 1: 23 – 20 = 3   or   28 – 20 = 8
Step 2: 3 + 28 = 31   8 + 23 = 31
Step 3: 31 × 2 = 3 (TENS of 23 and 28) = 62
Step 4: 3 × 8 = 3 (last digit of 23 and 28) = 24
Step 5:

H	T	U
6	2	
+	2	4
6	4	4

23 × 28 = 644

2   31 × 32

Step 1: 31 – 30 = 1 or   32 – 30 = 2
Step 2: 1 + 32 = 33   2 + 31 = 33
Step 3: 33 × 3 = (TENS of 31 and 32) = 99
Step 4: 1 × 2 (last digit of 31 and 32) = 2
Step 5: 31 × 32 = 992

3   62 × 69

Step 1: 62 – 60 = 2   or   69 – 60 = 9
Step 2: 2 + 69 = 71   9 + 62 = 71
Step 3: 71 × 6 = (TENS of 62 and 69) = 426
Step 4: 2 × 9 (last digit of 62 and 69) = 18
Step 5:

Th	H	T	U
4	2	6	
+		1	8
4	2	7	8

62 × 69 = 4 278

Limitation: You can only multiply numbers with the same TENS.

Base 100

This is quite similar to base 10, however, the reference number will be 100.

Numbers below 100

This can be applied to numbers as low as 50s. However, it is preferable to apply this technique to numbers close to 100 i.e. the 90s in order to avoid getting two digits numbers when you subtract from 100 which will make it cumbersome.

Steps

$A \times B$

1   Subtract both numbers from 100: $100 - A = C$

$$100 - B = C$$

2   Subtract the answer from the other number:

$100 - A = C$

$100 - B = D$

That is,

$B - C = E$

$A - D = E$

Answer obtained for both will be the same.

E will represent **HUNDREDS**.

3   Multiply the two values you got initially $C \times D = F$.

F will represent **TENS** and **UNIT**

4   Therefore, $A \times B = EF$

NOTE: E and F may be more than one-digit numbers.

Example 7.6

1   92 × 94

    Step 1: Subtract both numbers from 100 : 100 – 92 = 8
                                             100 – 94 = 6

    Step 2: Subtract the answer from the other number:
        100 – 92 = 8

        100 – 94 = 6
        That is,
        92 – 6 = 86
        94 – 8 = 86
        '86' will represent **HUNDREDS.**

    Step 3: 8 × 6 = 48 (**TENS** and **UNIT**)

    Step 4: 92 × 94 = 8 648

2   96 × 91

    Step 1: Subtract both numbers from 100 : 100 – 96 = 4
                                               100 – 91 = 9

    Step 2: Subtract the answer from the other number:
        100 – 96 = 4

        100 – 91 = 9
            That is,
        96 – 9 = 87
        91 – 4 = 87
        '87' will represent **HUNDREDS.**

    Step 3: 4 × 9 = 36 (**TENS** and **UNIT**)

    Step 4: 96 × 91 = 8 736

Alternative Method

1   Subtract both numbers from 100 : 100 − A = C

$$100 - B = D$$

2   Multiply C × D = EF

EF will represent **TENS** and **UNIT**

N.B: If EF gives one digit, place 0 in front e.g. 2 × 2 = 04

3   Add C + D = GH

4   Subtract from 100; 100 − GH = IJ

IJ will represent **HUNDREDS**

5   A × B = IJEF

To authenticate this method, we will solve the same examples previously solved to know if we will arrive at the same answers.

1   **92 × 94**

     Step 1: Subtract both numbers from 100 : 100 − 92 = 8

$$100 - 94 = 6$$

     Step 2: 8 × 6 = 48 (**TENS** and **UNIT**)

     Step 3: 8 + 6 = 14

     Step 4: 100 − 14 = 86 **HUNDREDS**

     Step 5: 92 × 94 = 8 648

2   **96 × 91**

     Step 1: Subtract both numbers from 100 : 100 − 96 = 4

$$100 - 91 = 9$$

     Step 2:   4 × 9 = 36 (**TENS** and **UNIT**)

     Step 3: 4 + 9 = 13

     Step 4: 100 − 13 = 87 **HUNDREDS**

     Step 5: 96 × 91 = 8 736

Since both methods gave the same answers, they can be used interchangeably.

Multiply the following base multiplication technique.

1 93 × 95   **2** 98 × 97   **3** 99 × 92

Numbers above 100

The process is similar to 'numbers above 10'. It is valid for 101 to 109. If the number obtained from multiplying the last digits of both numbers is one-digit, place 0 in front of the digit, e.g. 2 × 3 = 6 ▢ 06.

Steps

AB × CD

1 Subtract 100 from one of the numbers
   i.e. AB – 100 = E or CD – 100 = F

2 Add the result to the other number i.e. E – CD = G or F + AB = G
   G will represent **HUNDREDS**

3 Multiply the last digit of both numbers B × D = H
   H will represent **TENS** and **UNIT**

4 AB × CD = GH
   NOTE: G and H may not be one digit number.

Example 7.7

1 **105 × 101**

   Step 1: 105 – 100 = 5   or   101 – 100 = 1
   Step 2: 5 + 101 = 106   or   1 × 105 = 106 (**HUNDREDS**)
   Step 3: 5 × 1 = 05 (**TENS** and **UNIT**)
   Step 4: 105 × 101 = 10605

2 **103 × 108**

   Step 1: 108 – 100 = 8   or   103 – 100 = 3
   Step 2: 8 + 103 = 111   or   3 + 108 = 111 (**HUNDREDS**)
   Step 3: 3 × 8 = 24 (**TENS** and **UNIT**)
   Step 4: 103 × 108 = 11124

Using base multiplication method, multiply the following;

1 102 × 107   **2** 109 × 103   **3** 104 × 106

Base multiplication can be applied to higher base such as 1000. However, the major setback is that one can only multiply numbers with the same TENS , which will not always be the case in real-life, hence, a need for a more general technique.

Rainbow Multiplication

It is a more general approach to multiply as it is not restricted to numbers with same TENS. Rainbow multiplication entails splitting every number into two; every other digit(s) and the last. For instance,

Number	Other Digit(S)	Last Digit
98	9	8
264	26	4
126 349	12 634	9

Steps

AB × CD

Note that A and C may represent more than one digit.

1   Split each number into two i.e. the last digit and every other digit(s)
2   Multiply the other digit(s) of both numbers i.e. A × C = E (**HUNDREDS**)
3   Multiply the last digit of both numbers B × D = G (**UNIT**)
4   Multiply all numbers with rainbow arcs i.e.

AB × CD

(A × D) + (B × C) =F (**TENS**)

Therefore,

AB × CD=EFG

1   45 × 21

Step 1: Split each number 45, 21

Step 2: 4 × 2 = 8 **(HUNDREDS)**

Step 3: 5 × 1 = 5 **(UNIT)**

Step 4:

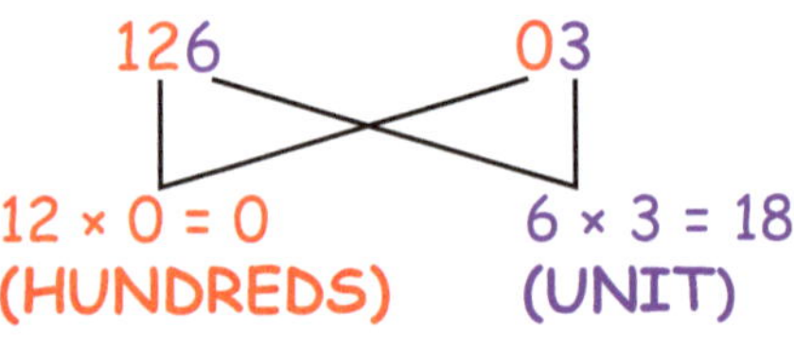

4 + 10 = 14 **(TENS)**

H	T	U
8		
+ 1	4	
		5
9	4	5

44 × 21 = 945

2   126 × 03

Step 1: Split each number 126, 03

Step 2: 12 × 0 = 0 **(HUNDREDS)**

Step 3: 6 × 3 = 18 **(UNIT)**

Step 4:

36 + 0 = 36 **(TENS)**

H	T	U
0		
3	6	
+	1	8
3	7	8

126 × 03 = 378

3   41 × 415

Step 1: Split each number 41, 415
Step 2: 4 × 41 = 164 (HUNDREDS)
Step 3: 1 × 5 = 5 (UNIT)
Step 4: 41   415

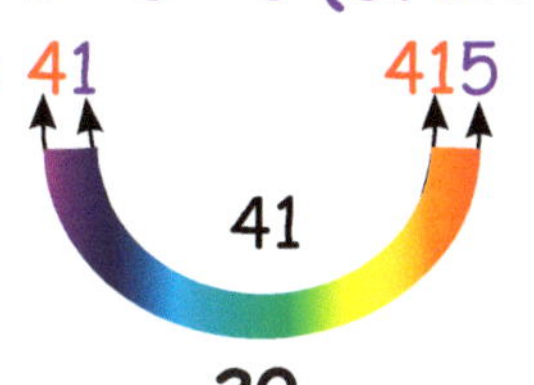

41

20

41 + 20 = 61 **(TENS)**

41   415

4 × 41 = 164   1 × 5 = 5
(HUNDREDS)   (UNIT)

TTh	Th	H	T	U
1	6	4		
		6	1	
+				5
1	7	0	1	5

41 × 415 = 17015

Do It Yourself 22

Using rainbow multiplication, solve the following

1   6 × 24
2   17 × 153
3   250 × 300
4   269 × 913
5   87 × 2681

Double-Half

Double-half technique can only be used if at least one of the numbers to be multiplied is an even number.

When multiplying two numbers and one of the numbers is even, divide the even number by two and multiply the other number by 2.

Do this over and over until the numbers left are very easy to multiply.

Example 7.9

1   **5 × 4**

4 is the even number.
Half 4 and double 5.
10 × 2
It is easy to tell the answer now.
10 × 2 = 20
Therefore,
5 × 4 = 20

2   **12 × 9**

12 is the even number.
Half 12 and double 9.
6 × 18
Half 6 and double 18.
3 × 36 = 108
Therefore,
12 ×9 = 108

3   **14 × 16**

Since both numbers are even, you can choose to half either of them.

Case 2:
Half 14 and double 16.
7 × 32
We cannot half 7.
7 × 32 = 224

Therefore,
14 × 16 = 224

4 72 × 13
72 is the even number.
Half 72 and double 13.
36 × 26
Half 36 and double 26.
18 × 52
Half 18 and double 52.
9 × 104
We cannot half 9:
9 × 104=936

Therefore,
72 × 13 = 936

Do It Yourself 23

With the aid of double-half technique, solve the following;

1 4 × 29   **2** 20 × 25   **3** 16 × 43

DIY 1

1. a) Tens
   b) Hundredths
   c) Thousands
   d) Ten Thousands
   e) Hundred Millions

2. a) 1- Ten Thousands,
      2- Thousands,
      1- Hundreds,
      6- Tens,
      2- Units
   b) 7- Ten Millions,
      6- Millions,
      1- Hundreds of Thousands,
      3- Ten Thousands,
      3- Thousands,
      5- Hundreds,
      4-Tens,
      3-Units,
      6-Tenths,
      1-Hundredths,
      1-Thousandths

3. a) 8- Thousands, 8- Tens
   b) 9- Ten Thousands,
      9- Tens,
      2- Thousands,
      2-Tenths,
      2-Hundredths

DIY 2

1. 48
2. 60

DIY 3

1. 49
2. 63

DIY 4

1. 24
2. 40

DIY 5

1. 72
2. 80

DIY 6

1. 27
2. 45
3. 54
4. 63
5. 81

DIY 7

1. 117
2. 126
3. 135
4. 153
5. 162
6. 171

DIY 8

1   70
2   80
3   90

DIY 9

1   12
2   14
3   72

DIY 10

1   432
2   3 705
3   750

DIY 11

1   81 972
2   130 845
3   252 109

DIY 12

1   42 067
2   152 184
3   252 109

DIY 13

1   1 920
2   109 720
3   942 619

DIY 14

1   1 066
2   283 590
3   12 282 873

DIY 15

1   144
2   22 221
3   36 519 912

DIY 16

1   4 224
2   10 6720

DIY 17

1   40
2   50
3   63

DIY 18

1   154
2   195
3   306

DIY 19

1   3 304
2   1 050
3   1 600

DIY 20

1   8 835
2   9 506
3   9 108

DIY 21

1   10 914
2   11 227
3   11 024

DIY 22

1	144
2	2 601
3	75 000
4	245 597
5	233 247

DIY 23

1	116
2	500
3	688

Index